ENVIRONMENTAL AUDITS

ENVIRONMENTAL AUDITS

Cliff VanGuilder

MERCURY LEARNING AND INFORMATION
Dulles, Virginia
Boston, Massachusetts
New Delhi

Publisher: David Pallai
MERCURY LEARNING AND INFORMATION
22841 Quicksilver Drive
Dulles, VA 20166
info@merclearning.com
www.merclearning.com
1-800-758-3756

This book is printed on acid-free paper.

Cliff VanGuilder *Environmental Audits*.

ISBN: 978-1-938549-60-1

Portions of this book have been quoted directly with permission from *Introduction to Environmental Science and Technology*. Dr. S. Amal Raj. Laxmi Publications Pvt. Ltd. 2008.

Library of Congress Control Number: 2013944482

141516321

Printed in the USA

Our titles are available for adoption, license, or bulk purchase by institutions, corporations, etc. For additional information, please contact the Customer Service Dept. at 1-800-758-3756 (toll free). Electronic versions of our titles can be found at *authorcloudware.com* and other sites.

CONTENTS

INTRODUCTION

T oday's environmental regulatory programs are incredibly complex, with each area of concentration taking up thousands of pages of detailed and often confusing laws, rules, regulations, and policy documents. This book is written to explain the definition of environmental audits, reasons for conducting these audits and provide detailed guidance on how to conduct them. It also breaks down these environmental programs into more understandable concepts, allowing the reader to know what the regulators seek for compliance.

DEFINITION OF ENVIRONMENTAL AUDIT

My personal definition of an environmental audit is: *"any documented activity to verify that environmental law(s), rule(s), or regulation(s) is/are in compliance at a particular location."*

I use this short definition for a reason—the variations on the complexity and level of effort involved in environmental audits have almost limitless possibilities. Environmental audits can vary from a simple phone call or field visit to verify compliance with one basic law, rule, regulation or company policy, to a very detailed, time-consuming (up to several weeks or months), and expensive marathon of planning, field work, report writing, and follow-up, all activities which will be described later.

An environmental audit can be as simple as inspecting a manhole, as illustrated in Figure I-1 below, or as complex as a comprehensive environmental audit of a vast, complex manufacturing complex, as illustrated in Figure I-2.

FIGURE I.1 *Courtesy of USEPA*

In its 1986 policy, The United States Environmental Protection Agency (USEPA) defines environmental auditing as "a systematic, documented, periodic, and objective review of facility operations and practices related to meeting environmental requirements."[1]

---

[1] United States Office of EPA 300-B-96-011 Environmental Protection Enforcement Agency and Compliance Assurance (2261A) Spring 1997 Page 1–3 http://www.epa.gov/compliance/resources/policies/incentives/auditing/envaudproguidemas.pdf

FIGURE I.2 *Courtesy of USEPA*

I prefer my simpler, yet broader, definition because the USEPA definition leaves no room for the occasional quick checks or periodic walk-through that are critical for every company to achieve consistent compliance. If every environmental audit met EPA's definition, very few audits would be conducted indeed.

SHOULD AN ENVIRONMENTAL AUDIT BE CONDUCTED?

Facilities face numerous decisions—few more important than whether or not to conduct an environmental audit. There are numerous compelling reasons to conduct environmental audits, and other, sometimes equally compelling reasons to avoid holding an audit. Some examples of good reasons to conduct an environmental audit include:

- as part of business decisions:
  - ensuring compliance is as economical as possible;
  - minimizing liability associated with non-compliance; and
  - ensuring business is viable and profitable with full environmental compliance.

- ▣ sale or purchase of facility(ies);
  - foreclosures;
  - acquisitions;
  - mergers;
- ▣ closing of facility(ies);
- ▣ required as a result of enforcement actions such as:
  - consent orders;
  - consent decrees; and
  - court orders

ADVANTAGES/DISADVANTAGES OF AN ENVIRONMENTAL AUDIT

There are many advantages to conducting an environmental audit, several of which are outlined below. A discussion of disadvantages to conducting environmental audits follows immediately after.

Advantages

- ▣ All facilities are required to follow all applicable environmental laws, rules, and regulations. Those that choose to ignore or avoid these standards and are audited or inspected by government regulators face serious consequences in the form of enforcement up to and including: notices of violation; consent orders; consent decrees; monetary penalties; and in the cases of criminal activity, even incarceration. For a recent example, on 12/16/2011, the USEPA released the following: "Texas Oil Company Sentenced to pay $12 Million for Clean Air Act Violations and Obstruction Crimes in Louisiana. Sentence is the largest ever criminal fine in Louisiana for air pollution. Pelican Refining LLC was was sentenced to pay a $12 million penalty, which includes a $10 million criminal fine and $2 million in community service payments that will go toward various environmental projects in Louisiana, including air pollution monitoring. The vice-president who oversaw operations at the Lake Charles refinery since 2005 from an office in Houston, Texas pleaded guilty on July 6, 2011, to negligently placing persons in imminent danger of death and serious bodily injury as a result of negligent releases at the refinery. Hamilton faces up to one year in prison and a $200,000 fine for each of the two Clean Air Act counts."[2]

- In addition to avoiding violations and penalties, environmental audits are now generally recognized as wise investments for businesses that desire success in today's uncertain economic and environmentally-conscious times. They can save the facility substantial money. I was personally involved in several regulatory audits wherein the facility realized savings. At the end of the Introduction, I have included a case study wherein a company I inspected discovered thousands of dollars in savings monthly as a result of my regulatory audit of their hazardous waste program.

- An accurate and comprehensive environmental audit, followed by complete implementation of any and all corrective measures, is very much like an insurance policy, ensuring that inspections by regulatory agencies will not result in costly enforcement actions.

- Accurate and comprehensive audits reduce liability and save the potential costs of penalties, fines, lawyer's fees, and lost production time while corrections are being made. As illustrated earlier, serious environmental crimes can also result in jail time.

- Good environmental compliance can improve the facility's public image, causing less friction with both regulators and the surrounding community.

- Staff involved in audits will have better environmental awareness, improving morale and staff ownership and participation in compliance.

- If an audit reveals problems or violations and they are corrected, potential environmental damage and/or personal injuries from these concerns might be prevented.

Disadvantages

- Environmental audits can be expensive. Some audits can be conducted using existing staff, but consultants are sometimes necessary to provide expertise not available in-house, along with possible sophisticated

---

[2] United States Environmental Protection Agency; News Releases by Date http://yosemite.epa.gov/opa/admpress.nsf/0/224A2AF1EC8927C7852579670063837A

sampling and laboratory analysis. Audits included as part of an enforcement action often require the use of an unbiased, independent contractor, pre-approved by the court or regulatory agency. These audits and the associated enforcement actions usually require the use of an attorney as well.

- Environmental audits can be time-consuming. Staff time is required to conduct in-house audits, tying up resources that would otherwise be able to accomplish other work. Even if an audit is conducted by an outside, independent contractor, staff still need to cooperate with the consultant, and give their time to answer questions and oversee the work of the consultant.

- Plant operations may be disrupted in many inspections, particularly for air, water, and hazardous waste pollution control equipment (trial burns, calibration of equipment and monitors, sampling, and analysis)

- An environmental audit can expose a company's liability:

  - Any deficiencies found during an audit must be corrected.

  - In addition, a problem identified during an audit may require reporting;

  - Finally, if audit recommendations are cost-prohibitive, and regulators discover the recommended corrections were not made in a timely fashion, there could be serious violations.

- Increased liability from pollution related incidents or injuries. If environmental damage or personal injury results from failure to conduct an audit or to act on audit recommendations, the company might be found liable for those damages.

NOTE

The final listed disadvantage above could be offset or even be seen as an advantage, because corrections from an audit might very well prevent future possible incidents of pollution (releases) or injuries from those incidents.

HISTORY OF ENVIRONMENTAL AUDITING

Environmental auditing started soon after the environmental awareness movement was born, when manmade sanitary and industrial pollution were recognized as dangerous to human health and the environment. Perhaps

FIGURE I.3 Love Canal Hazardous Waste Cleanup Site, Niagara Falls, NY. *Courtesy USEPA*

the most highly recognized event that created the environmental movement was the discovery of the Love Canal hazardous waste dump site in Niagara Falls, New York.

The Love Canal hazardous waste site made national and international news in the early 1970s when residents living near the dump site realized they were experiencing strange odors, and noxious liquids from the abandoned dump site were oozing into their basements. For a more detailed description of the events leading up to the discovery of Love Canal, the ensuing investigation, and cleanup of the Love Canal Hazardous Waste site in my text, *Hazardous Waste Management, an Introduction.*[3] (Shameless plug.)

The United States Protection Agency (USEPA) was formed by Congress in 1970 to regulate pollution, and was closely partnered with the New York State Department of Environmental Conservation (NYSDEC) in the Love Canal investigation and cleanup.

---

[3] *Hazardous Waste Management, an Introduction,* **ISBN:** *781936420261* by Cliff VanGuilder http://www.merclearning.com/titles/hazardous_waste_management.html

On July 9, 1986, the USEPA announced the "Environmental Auditing Policy Statement" (51 CFR 5004) (1986 audit policy), updated December 1988 to EPA-305-B-009[4] and followed that with the "Incentives for Self-Policing: Discovery, Disclosure, 1-2 Correction and Prevention of Violations" on December 22, 1995 (60 CFR 6706) (1995 or final audit or self-policing policy).[5]

The 1986 audit policy states that "it is EPA policy to encourage the use of environmental auditing by regulated industries to help achieve and maintain compliance with environmental laws and regulation, as well as to help identify and correct unregulated environmental hazards."

Many states have also developed and issued their own environmental audit policies. For example, on August 12, 1999, The New York State Department of Environmental Conservation instituted Commissioner Policy CP-19, entitled "Small Business Self-Disclosure Policy."[6]

The purpose of this policy is set forth as follows: "This document sets forth DEC's Policy for promoting environmental protection and improving compliance rates by establishing a process for adjusting penalties where small businesses detect, voluntarily disclose and expeditiously correct certain violations discovered through environmental audits or compliance assistance. This Policy will provide the regulated community with greater certainty regarding the Department's response to self-disclosed violations. This enhanced certainty will reduce the fear associated with reporting violations and thereby foster compliance auditing and compliance assistance."

The USEPA policy, along with reinforcing state policies, is clearly written to encourage industries to conduct environmental audits without fear of reprisal.

---

[4] USEPA Environmental Auditing Policy Statement" (51 CFR 5004), updated December 1988 to EPA-305-B-009 at http://infohouse.p2ric.org/ref/43/42045.pdf

[5] Incentives for Self-Policing: Discovery, Disclosure, 1-2 Correction and Prevention of Violations, December 22, 1995 (60 CFR 6706) (1995 or final audit or self-policing policy) at http://www.epa.gov/compliance/resources/policies/incentives/auditing/finalpolstate.pdf

[6] The DEC Policy System, Department ID: CP – 19, Issuing Authority: John P. Cahill, at http://www.dec.ny.gov/regulations/25246.html

This textbook will outline the entire environmental audit process, including:

- determining the necessary and proper scope of work (which particular and how many environmental programs need to be audited);

- planning an audit, including selecting the resources to conduct the audit;

- selecting who should conduct the audit;

- conducting the audit;

- reviewing the findings from the audit;

- making decisions on whether to take corrective actions recommended by the audit; and

- reviewing and revising business plans as needed.

THREE UNIQUE FEATURES OF THIS BOOK

There are three features of this book that make it different from the other books on environmental auditing:

1. Comprehensive coverage of important, lesser publicized environmental programs.

Many books on environmental audits only deal with the major environmental programs. In addition to a complete guide on auditing the major, more popular programs of:

- Air Resources;

- Solid and Hazardous Waste;

- Water Resources; and

- Environmental Cleanup,

This book includes a comprehensive guide to lesser know but very important environmental issues faced by businesses today, including sections on:

- Chemical/Petroleum Bulk Storage;

- Radioactive waste materials;

- Storm water;

- Pesticides;

- Wetlands;

- Stream Disturbance;

- Lands and Forests;

- Mining (mined land reclamation); and

- Wildlife.

Within these chapters are detailed sections within little known and sometimes overlooked requirements. For instance, in addition to the standard air resources regulatory requirements, the author has included sections on

- Open Burning of Household Waste;

- Waste Oil Space Heaters; and

- Idling of Heavy-Duty, On-Road Vehicles.

2. Guidance on Leadership in Energy and Environmental Design (LEED)

This book also includes a chapter on Leadership in Energy and Environmental Design (LEED), a program developed for the design of buildings with reduced energy usage and enhanced environmental compliance.

The voluntary LEED Program was developed in 1998 to provide building owners and operators an accurate, concise, and understandable framework for finding and implementing practical and measurable design, construction, operations and maintenance solutions for "green" buildings.

3. Chapters on household and business office audits

This book includes chapters for household and office environmental reviews. All homeowners and businesses would benefit from conducting environmental audits of their homes and offices, in that they would be able to identify the hazards within these places, such as: dangerous chemicals; asbestos; radon; and other hazards. These audits provide safer homes and workplaces. In the case of offices, these audits will help avoid notices of violations for improper disposal of fluorescent lamps, hazardous cleaners, paints and other hazardous wastes.

Environmental audits by homeowners is a new concept that is growing in popularity. The environmental awareness of homeowners and office

FIGURE I.4 *Courtesy of USEPA*

managers is improving rapidly, with the resources available on the internet, medical shows, and social media. Not everyone is aware of the environmental hazards in our own homes, but these dangers are real and can easily be identified and corrected at little or no cost for most issues.

Why does a home audit make sense?

Homeowners are unknowingly exposed to dangerous environmental toxins in their homes every day. The federal, state and local governments do not regulate the chemicals that are used for cleaning households. Some of these chemicals are highly toxic to humans and animals, and require special storage and handling instructions, including the use of personal protective equipment (PPE) like respiratory protection, goggles, splash aprons, and chemical-resistant gloves if used in an industrial setting.

I have written an e-book summarizing the hazardous materials commonly found in homes and offices. The book is entitled *Detox Your Home Taming the Toxic Menace*.[7] An electronic copy of this book is included as a supplement to this textbook. The purpose of this supplement is to educate everyone on the potential hazards people may encounter in their own

---

[7] *Detox Your Home—Taming the Toxic Menace*, copyright registration # TXu 1-671-612 at http://www.amazon.com/Taming-Toxic-Menace-Your-ebook/dp/B008G65RWQ

homes, and offer ways to reduce or remove these hazards by using safer, environmentally friendly alternatives.

This book provides a comprehensive guide to conducting environmental audits, no matter what type of business is being reviewed. If these audits are conducted accurately and thoroughly, they will provide these businesses (and homeowners) with the assurance they can respond to regulatory inspections or citizen inquiries, and as an added bonus, provide a safer place to work and live.

Business office audits

Business offices can generate a surprisingly large number and types of hazardous wastes, from fluorescent lamps, to mercury containing thermostats, aerosol cans, batteries, waste paints, commercial office cleaners, pesticides, and all types of electronic wastes. All business offices are subject to hazardous waste inspections, and if the wastes are not being managed in accordance with the regulatory standards, may be subject to enforcement actions.

What is not included in this book

This book does not include sections on health and safety requirements primarily governed by the federal Occupational Safety and Health Administration (OSHA), regulating workplace safety issues, such as exposure to lead, asbestos, workplace injuries, and health and safety training.

Issues such as lead and asbestos pollution can bridge over into any of the pollution control programs; air, solid waste, or water, but workers' exposure to these pollutants is covered under OSHA.

GOVERNMENTAL PERSPECTIVE

When it was formed in 1970, the federal United States Environmental Protection Agency (USEPA) was charged by Congress to promulgate rules and regulations based on all environmental laws. In addition, the USEPA was directed to enforce all of these environmental laws, and the policies, rules, and regulations that were developed from these laws. Obviously, part of EPA's enforcement strategy encompassed environmental compliance inspections, which are, in fact, program-specific environmental audits.

The USEPA uses regulatory checklists for all types of environmental regulatory inspections. These checklists are available from state and federal regulatory agencies, and are excellent resources for developing environmental audit checklists.

In order to accomplish the enforcement of these standards, the USEPA set up a regulatory and enforcement delegation program available to each of the 50 states, and Washington D.C., Puerto Rico, Guam, American Samoa, and the Northern Marianas. These governmental entities were all offered funding by the USEPA, provided they developed and promulgated environmental regulations at least as stringent as the federal regulations. They were also required to set up permitting and enforcement programs to administer these programs. Some of the states and territories applied for this delegation, while others elected to have EPA run their programs.

The result of these delegation agreements was state and territorial compliance inspection programs that required to be at least as stringent as the USEPA regulations. These regulations are used for conducting compliance inspections for the three major environmental regulatory programs (air, water, solid and hazardous materials), and later radiation, pesticides, etc.

Environmental compliance inspections are most often conducted by the state or territorial inspectors, but occasionally a USEPA inspector might conduct inspections, either because a state or territory has not been delegated authority for that inspection, or because the USEPA is conducting an independent or oversight inspection to make sure the state or territorial inspector is doing inspections correctly.

The USPEA developed guidance for compliance assurance at federal facilities in a document entitled *Environmental Protection Enforcement Agency and Compliance Assurance EPA 300-B-96-011 (2261A) Spring 1997*[8]

AUTHOR'S PERSPECTIVE

As a former governmental employee, retired after 35 years of regulating facilities that managed solid waste, hazardous waste, water, air and

---

[8] United States Environmental Protection Agency, *Environmental Protection Enforcement Agency and Compliance Assurance EPA 300-B-96-011 (2261A) Spring 1997* at http://www.epa.gov/compliance/resources/policies/incentives/auditing/envaudproguidemas.pdf

pesticides, I would like to offer my perspective on environmental audits conducted by government staff.

It might surprise some people to hear that most of the regulatory enforcement staff that I knew at all levels of government wanted facilities to successfully comply with all laws, rules and regulations. While it is true there may be a few inspectors who enjoy finding violations, it was my experience that these inspectors were very much in the minority. When I conducted inspections and supervised inspection staff in a state environmental regulatory setting, it was to help facilities understand and comply with the very complex rules faced every day by these companies. That required: personally learning these regulations; conducting seminars to explain the regulations to industries; going on hundreds of inspections to ensure compliance; supervising and reviewing thousands more; and making sure we used those inspections as outreach events to make industries aware of their requirements and help them identify the most efficient way to comply with those requirements.

For the most part, regulatory staff and inspectors I knew would rather receive a phone call or a visit from facility representatives with questions on how to comply with the regulations than to have the facility attempt to interpret the rules themselves. The facilities that interpreted the regulations without the benefit of knowing what the inspectors wanted were often found to be in violation at their next inspection. A very experienced Environmental, Health, and Safety (EHS) coordinator once at a regulated facility told me; "I care most about the inspector's interpretation of the rules because those are the standards I must meet." This is a very insightful observation about the psyche of the regulatory inspector. Inspectors are trying to ensure compliance with the laws, rules, and regulations as they understand them, and any variance from that understanding might be a violation. If the facility representative understands the inspector's understanding of the rules, inspections can be much more successful than when there is disagreement.

That is not to say the inspector's interpretation is always correct. Inspectors are human and may interpret the rules in an inconsistent fashion. The inspector may have misunderstood their instruction, or the instruction may have been incorrect.

When a disagreement about an interpretation of the laws, rules, or regulations occurs, the facility representative would be wise to ask the inspector the basis for the interpretation, and if after that explanation, the facility representative and inspector still disagree, the facility representative can ask the inspector to call his/her supervisor or regulatory expert to verify the inspector's interpretation. If this check still indicates an unfair or inconsistent interpretation, the facility representative can appeal the issue even higher.

At each point of disagreement and level of appeal, the facility representative needs to decide if the disagreement is worth the effort, since the inspector believes their position to be correct. The inspector is trying to enforce compliance with the laws, rules, and regulations, and has the power to cite the facility whether their interpretation is correct or not.

If the cost of making the correction is less than the potential cost of enforcement, it is recommended that the facility comply with the inspector's standard. At times, in addition to cost, the facility owners/managers need to also consider the issues of continuing compliance, precedent, and the impact on the facility as a whole. Any points of disagreement can still be appealed after the inspection and correction, but the burden of responding to an enforcement action can be averted.

If the cost of complying with the inspector is greater than the potential cost of enforcement, the facility can accept the violation and appeal it during the enforcement phase. There is generally plenty of time to appeal the violation, but the alleged violation is public record and customers can review it if requested through the proper channels.

As mentioned earlier, these rules and regulations can be very confusing, but careful review of legislative history, regulatory preambles, policy memos and the regulations themselves can provide insight to the intent of the rules and regulations. Pointing out the language that supports your position can often help sway the regulatory staff to agree with your position.

It helps to always be prepared for an inspection. Chapter 10 of this book includes a segment suggesting what to do when a regulatory inspector visits your facility.

INDUSTRIAL PERSPECTIVE

Owners and managers of facilities make decisions on how to run their business on a continual basis. Many of those decisions concern the priority of compliance with all of the laws, rules, and regulations that apply to their business. Compliance with these standards almost always cost the facility time and money, and these costs must be factored into the priority of where to spend time and other resources.

Is Environmental Compliance a Priority?

It should come as no surprise that the facilities who decide to make environmental compliance a high priority almost always pass compliance audits with few or no substantial violations. Facilities who choose to make environmental compliance a lower priority are more likely to have more numerous and serious violations, depending on the amount resources withheld from compliance.

This is not to say that facilities with a high priority on environmental compliance always pass their compliance inspections. If the facility's staff is not properly trained in compliance, and/or does not seek advice from the government regulatory staff, their interpretation of the rules could lead to deviations from government inspectors' understanding of the standards, causing potential violations.

Ironically, if a facility decides to make environmental compliance a lower priority and is found to be in serious violation of the rules and regulations, part of the enforcement process is to require the facility to pay for an environmental audit by an independent consultant at a much higher cost, and to pay for the corrections recommended by the independent consultant. From my observations, facilities that failed to make environmental compliance a priority or purposely avoided costs by not complying with the rules always ended up spending more money defending themselves in hearings or court for their violations, and paying for the corrections of their violations than if they had complied in the first place.

On the negative side, hiring and maintaining capable environmental staff presents a supervisory and cost challenge to management.

Companies that neglect to account for all costs in their business plan face financial losses and potential failure. Environmental compliance costs are a very important part of the expense side of the ledger and are growing at an unpredictable rate.

Case Study—Company Saves Money Because of a Regulatory Inspection

I performed an unannounced hazardous regulatory inspection at a manufacturing facility before I retired from the state agency where I previously worked. When the secretary called the facility manager to announce my presence, he sent me his Information Technology Specialist (ITS), a person who had very limited knowledge of the hazardous waste program (a very dangerous move as discussed in Chapter 2). The company representative acted very nervous and asked me if I had a warrant for the inspection. I explained that it was a routine inspection, but if he required a warrant I could get one fairly quickly. I also explained that if I needed to get a warrant, I would likely return with several inspectors because asking for a warrant implied something might be wrong at the facility. After hearing this, he consented to show me the hazardous waste storage and handling areas, and other than a few minor issues, everything was in substantial compliance. However, as the company representative became more relaxed and opened up about the waste management fees from the vendor that took their waste away, it became clear to me the facility was paying far more for waste disposal than they needed to pay. Their vendor was identifying much of their waste hazardous waste, and charging the facility several times more for the waste disposal than normal. In addition, identifying the wastes as hazardous waste required hazardous waste manifests and land disposal restriction forms, adding still more charges. After the company representative contacted the vendor, he called me back and told me the changes in waste identification had saved the company over $2,000 per month. They had fired the original vendor and hired a company that was helping them find further savings. After that incident, I received regular calls from the company representative and eventually the facility manager asking for more regulatory advice. The manager and the company representative were delighted I had inspected them, and welcomed me back any time.

Businesses face difficult decisions, with perhaps the most critical being whether part or all of the enterprise is economically viable (can make sustainable profits). The costs of complying with all environmental laws, rules, and regulations have become an ever-increasing part of the cost calculations, and should receive consideration and be included in the operating budget of every business plan. Failure to calculate the costs of environmental compliance and factoring them into the business plan could cause any business to struggle or fail.

In the author's experience, companies that recognized environmental compliance as a high priority generally had a higher success rate than those who did not.

Final note: One very important tool used by regulators interpreting rules and regulations are documents used to reveal the intent of the governmental body that authored the legislation. In addition, government has developed official regulatory guidance which must be developed, reviewed and approved through a prescribed process before issuance.

The legislative history for all federal laws is contained in the Library of Congress, available online at http://www.loc.gov/law/help/leghist.php[9]. The USEPA and State regulatory preambles often reveal some of this information, but more accurate information concerning legislative intent is contained in the law's legislative history. This history can be in many forms for state and local governments, including bill memos, white papers, and other forms of correspondence prepared by legislators or their staff.

The people preparing the language of the regulations rely on the legislation, the background documents, and the regulatory preambles, but there is often room for interpretation of these documents that are different from the rules and regulations written by staff.

Anyone questioning the interpretation of rules and regulations should always study the legislative history, regulatory preambles, and official regulatory guidance before accepting the language of the rules and regulations. This research may include reviewing records at federal, state and local archives, either in person or online, and may also directly contacting legislative staff for interpretation of the sponsors and lawmakers when the law was drafted. Many regulatory staff also have background on the environmental and legal basis for regulations, and can often provide helpful insights in these areas.

---

[9] Website for US Library of Congress at http://www.loc.gov/law/help/leghist.php

CHAPTER 1

SUMMARY OF ENVIRONMENTAL PROGRAMS AND REGULATIONS

As mentioned in the Introduction, one of the unique features of this text is the recognition, categorization, and explanation of all of the most well-known, and many of the less-known, environmental programs. Care is taken to explain these programs in sufficient detail to allow the auditor(s) or manager(s) to understand how environmental programs may or may not be regulated. This is in contrast to other audit texts that spend the majority of pages explaining the audit process without explaining the basic requirements of each environmental program (and without including descriptions of many environmental programs).

This text devotes several chapters to explain the audit process. Before informing the reader about how to plan, conduct, and write up an audit in Chapters 2–9, this chapter lays out all of the various environmental programs and their basic tenets.

For the sake of organization, these programs are segregated into major (most publicized), medium (less publicized), and minor (least publicized) environmental categories in this chapter. Other recent environmental audit books address what one would call the major programs, but this text covers virtually all environmental programs.

1.1 MAJOR ENVIRONMENTAL PROGRAMS

Environmental regulations are categorized into three major areas:

1. Solid and hazardous wastes

2. Water resources

3. Air resources

1.1.1 Solid and Hazardous Wastes

Solid waste and hazardous waste are listed together mainly for regulatory reasons. Hazardous waste is considered a subset of solid waste by federal law. Radioactive wastes can be mixed with hazardous waste, and may be regulated by U.S. Environmental Protection Agency hazardous waste regulations, U.S. Department of Energy regulations, U.S. Nuclear Regulatory Commission or any combination of these three.

1.1.1.1 Solid wastes

Solid wastes are any discarded (abandoned, recycled, or considered waste-like) materials. Because this is a regulatory description instead of a scientific description, solid wastes do not necessarily need to be in the solid phase. They can also be liquid, semi-solid, or containerized gaseous material.

Examples of solid wastes that may be encountered include, but are not limited to, the following:

- Domestic refuse (garbage)
- Construction and demolition debris, asbestos
- Waste tires
- Scrap metal
- Uncontaminated used oil
- Latex paints
- Empty aerosol cans, paint cans, and compressed gas cylinders
- Antifreeze (a potential hazardous waste)

Solid wastes must be disposed in conformance with federal, state, and applicable local laws, rules, and regulations (the federal regulations are

FIGURE 1.1 Landfill Compactor (Town of Colonie, NY).

under 40 CFR Part 257 and 258). The regulations allow for, and encourage, recycling and reclaiming of some wastes, but in general, solid wastes must be disposed in permitted facilities.

In many states, each municipality, township, parish, and/or county may have their own more specific requirements for waste disposal and recycling. Facilities should be sure to check with all levels of governmental bodies. Complete liability for solid waste management lies with the waste generator and the waste hauler collecting and consolidating the waste for disposal.

The best way to prevent unlawful disposal in Dumpsters (roll-offs) is by keeping them in secure areas (fenced) and conspicuously posting signs directing which specific materials are acceptable for disposal and where unacceptable wastes should be disposed.

Facilities may need a solid waste permit or registration if they want to treat or dispose of any solid waste on their property. It is highly advisable to check with the appropriate governmental authority before taking any action.

In many states, waste hauler permits are generally required to transport solid waste from a facility's property. In some states, a state waste hauler permit is not required if the facility transports its own solid waste (i.e. Wisconsin, Michigan, and New Jersey).

1.1.1.2 Hazardous wastes

As mentioned earlier, federal laws, rules, and regulations consider hazardous wastes a regulatory subset of solid wastes which either:

- Exhibit a hazardous characteristic, including

  - Ignitable (flashpoint <140° F, ignitable solids, oxidizers, and compressed gases)

  - Reactive (explodes when exposed to air or water, cyanides, sulfides, materials capable of detonation or explosion, materials that undergo violent change or reaction, or materials that generate toxic gases)

  - Corrosive (pH less than 2 or greater than 12.5)

- Contain certain contaminants above a certain level (toxic)

- Are listed in Title 40 of the Code of Federal Regulations (40 CFR) Part 261

Hazardous waste management requires proper shipping by a licensed transporter using a hazardous waste manifest followed by specific processing and a defined ultimate disposal site (usually not a solid waste disposal site). All facilities are subject to all hazardous waste laws, rules, and regulations, including commercial businesses, governmental organizations, schools, and other entities.

Examples of hazardous wastes that may be encountered in typical businesses include, but are not limited to, the following:

- Fluorescent lamps, mercury vapor lamps, HID lamps, and smoke detectors

- Many oil-based paints

- Latex paints manufactured prior to August 1990 (due to solvent and heavy metals content)

- Lead/acid batteries in storage until recycled

- Rechargeable batteries (nickel-cadmium, lithium, silver oxides, mercury, and lead)

- Used oil contaminated with hazardous waste (gasoline, solvents, heavy metals, pesticides, etc.)

- Spent or unusable waste-cleaning products

- Compressed gases that are at a pressure above standard pressure and temperature that are flammable or will support combustion

- Contaminated rags before recycling/laundry

- Transformers containing polychlorinated biphenyls (PCBs) greater than 50 parts per million (ppm)

- Certain used, expired, or unregistered pesticides

- Used electronic equipment (computers, monitors, etc.)

- Spent parts-cleaner fluids, except certain non-halogenated parts washers with a flash point greater than 140° F

For the purposes of hazardous waste definition, chlorine is the most commonly encountered halogen. However, a halogen can be any one of the following chemically related elements: chlorine (Cl), bromine (Br), fluorine (F), iodine (I), and astatine (At).

To help identify whether a waste is hazardous, the auditor should double-check labels and material safety data sheets (MSDS) when possible, and determine how the waste was generated.

Common Labeling Requirements for Containers of Hazardous Waste

Some common labeling requirements for containers holding hazardous waste include the following measures to ensure compliance with federal, state, and local laws, rules, and regulations:

- Each container must be marked with the words "Hazardous Waste" and with other words to identify the contents.

- The date upon which waste accumulation begins must be clearly marked and visible for inspection on each container.

- Containers are no longer subject to hazardous waste laws, rules, and regulations if they are empty. The containers are considered empty if

- All wastes have been removed using the practices commonly used to empty that container

  - No more than 1 inch of residue remains in the bottom of the container or no more than 0.3 percent by weight of total capacity of the container remains for containers less than or equal to 110 gallons

  - No more than 0.3 percent by weight of total capacity of the container remains for containers greater than 110 gallons

  - When the pressure in hazardous-waste compressed-gas containers approaches (is essentially equal to) atmospheric pressure

Rules for Safe Storage and Handling of Hazardous Wastes and Chemical Products

The following rules apply for the safe management of hazardous wastes and chemical products:

Hazardous Waste
FEDERAL LAW PROHIBITS IMPROPER DISPOSAL
If found, contact the nearest police or public safety authority, and the
Washington State Department of Ecology or the Environmental Protection Agency

Accumulation Start Date:	Generator Name:
Reportable Quantities (RQ): lbs	Address:
40 CFR Subchapter J, Part 302, Table 302.4	City:
Manifest Document #:	State:
Emergency Response Guide #:	Zip:
EPA Waste Code(s) and/or Characteristic(s)	EPA ID #:

EPA/DOT Shipping Name:

  Hazard Class:

  UN/NA #:

  Packing Group (PG):

In the event of a spill or release of this hazardous waste, contact the US Coast Guard
National Response Center at 1-800-424-8802 for information and assistance.

FIGURE 1.2 Hazardous Waste Label. (*Courtesy USEPA*)

- Keep products in original containers or properly label new ones.

- Check containers frequently for signs of leaks, spills, and deterioration.

- Make sure containers are kept closed, in good condition, and are made of compatible materials.

- Never mix different chemical products (or wastes) unless directed to do so.

- Store away from extreme heat or cold and away from human or animal contact.

- Never dispose of chemical products and wastes in septic systems.

- Use personal protective equipment, if necessary, when handling hazardous wastes, as prescribed by the MSDS

Following is a photograph of a hazardous waste container that held incompatible hazardous materials.

FIGURE 1.3 Bulging Steel Drum. *From Investigative Report NO. 2002-02-I-NY, Figure 4, Page 20, U.S. Chemical Safety and Hazard Investigation Board, September 2003.*

1.1.1.3 Universal wastes

Universal wastes are certain types of hazardous wastes that are generated on a regular basis by almost every business and industry (universally). Many of these wastes are so commonly generated that their associated hazards and dangers are often downplayed or ignored. Streamlined rules and regulations apply to the following, provided they are still in their original cases: batteries, pesticides, thermostats, and lamps.

A universal waste generator can still qualify as a conditionally exempt small quantity generator if the only hazardous wastes they generate are universal wastes and they manage those wastes as universal wastes.

Following is a photograph of a container holding universal waste fluorescent lamp ballasts. Note the relaxed labeling standards (no hazardous waste label required).

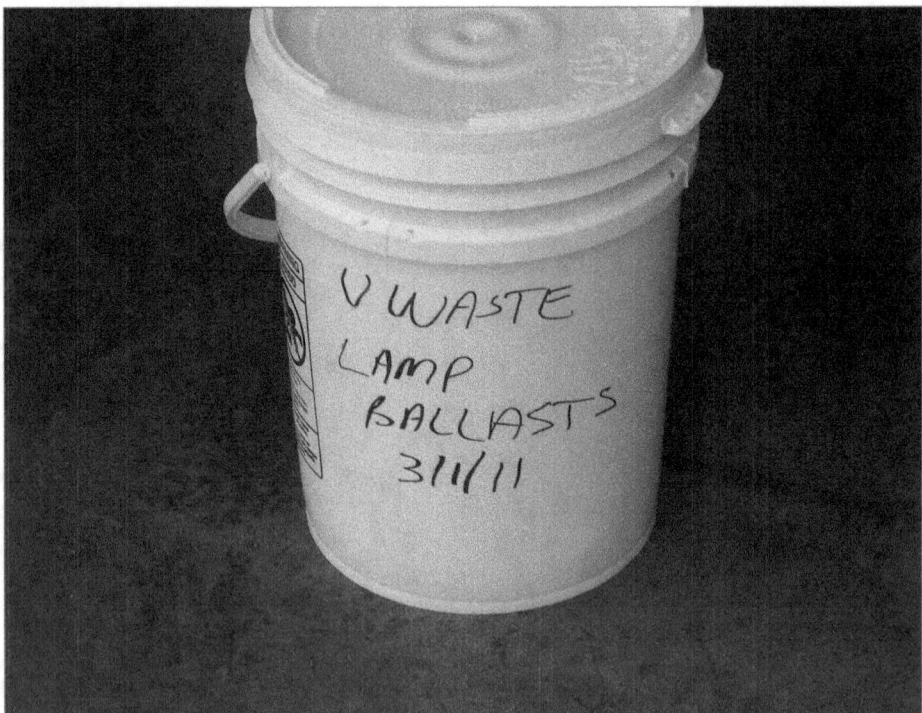

FIGURE 1.4 A container holding universal waste fluorescent lamp ballasts. *Courtesy of Faith Baptist Church, Rexford, NY.*

1.1.1.4 Radioactive wastes

Radioactive wastes are produced by activities that manufacture or use radioactive materials. Examples are:

- Mining

- Nuclear power generation

- Various processes (from industry, defense, scientific research, and medicine) generating by-products that include radioactive waste

Radioactive wastes can be in gas, liquid, or solid form. Their levels of radioactivity can vary, and these wastes can remain radioactive for a few hours or several months or even hundreds of thousands of years.

Radioactive wastes are regulated by several agencies, including the United States Nuclear Regulatory Commission (USNRC), the United States Department of Energy (USDOE), the United States Environmental Protection Agency (USEPA), and by most states.

The currently accepted radioactive labels with yellow, magenta, and red colors are being replaced with a new symbol.

The newly developed universal radioactive waste label is shown in Figure 1.5.

The most common radioactive waste encountered is low-level radioactive waste, which is regulated by most state agencies and overseen by the USEPA and the USNRC. Because most of the facilities are regulated by several regulatory agencies, environmental audits of these types of waste can be very detailed and complex. Detailed guidance on conducting radioactive waste management audits is available on the USNRC and USEPA Websites.

FIGURE 1.5 The new universal radioactive waste label. *Courtesy of USEPA*

1.1.2 Water Resources

The USEPA regulates wastewater treatment and discharges under the Clean Water Act, requiring permits for certain activities under the National Pollutant Discharge Elimination System (NPDES). These permits may be required for any of the following activities:

- Industrial pretreatment
- Concentrated animal feeding operations (CAFOs)
- Pesticide discharges
- Combined sewer overflows (CSOs)
- Sanitary sewer overflows (SSOs)
- Storm water
- Whole effluent toxicity (WET) for aggregate toxic effect

FIGURE 1.6 Inspecting Waste Water Treatment. *Courtesy USEPA*

1.1.2.1 Vehicle washing

Outside washing of vehicles with plain water is generally acceptable, provided there are no detergents, even those labeled biodegradable. To prevent adversely affecting water quality, vehicle washing should be done in a manner that minimizes and directs runoff away from any surface waters or storm sewers that drain to surface waters. Use of a high-pressure wand is recommended to reduce water usage. Detergent washes of vehicles are acceptable if done inside a building that has floor drains which discharge to either a municipal sewer system or to a holding tank. If drains discharge to an oil/water separator, detergents should not be used because they can affect the performance of the separator.

1.1.2.2 Floor drains

Any buildings with floor drains should be identified and evaluated as part of the environmental audit process. Floor drains that discharge to a holding tank or municipal sewer system may be subject to pretreatment requirements of a municipal sewer or publicly owned treatment works. Floor drains that do not discharge to a holding tank or municipal sewer system should be plugged until they can be evaluated to ensure that the discharge is properly managed.

1.1.2.3 Septic tank/disposal system maintenance

Septic tanks should be inspected and pumped out every two years. The tanks must be pumped by a licensed hauler. Do not use additives that claim to clean the tank. Do not dispose of waste other than sanitary waste in a septic system.

1.1.2.4 Oil/water separator maintenance

Oil/water separators should be inspected periodically for oil and solids buildup. The frequency of inspection depends on how much the separator is used. The separator must be cleaned out by a licensed hauler.

FIGURE 1.7 Properly Plugged Floor Drain. *Courtesy USEPA*

1.1.2.5 General permit activities

Depending on the location, a permit may be needed for any of the following:

- Construction in a flood-hazard area
- Work in waters that are or affect a public water supply
- Work that results in a discharge of wastewater or storm water
- Construction and maintenance of dams
- Excavation or filling of wetlands
- Construction resulting in disturbance of one acre or more

A federal 404 United States Army Corps of Engineers (USACOE) permit may be required for any of these activities. While a facility may not need a state protection of waters permit, it may still need a water quality certification from the local regulatory office.

Even if no permits are needed, a facility will still need to adhere to all water-quality standards and prevent discharges of sediment, oil, grease, etc. into a waterway or a wetland from a job site. Before conducting any of these activities, the facility should check with its local environmental regulatory authority.

1.1.3 Air Resources

The USEPA regulates air discharges under the Clean Air Act, first signed in 1970, and amended in 1990.

Each individual state has been delegated by the USEPA to administer and issue or deny companies air permits and registrations. The states are also responsible for collecting and administering regulatory fees for these permits (these fees are the funding for states to administer the program).

FIGURE 1.8 Air Discharges. *Courtesy USEPA*

Each facility and projects within the facility need to be evaluated to see if a project is classified as a major permit (requiring a Title V permit), a minor permit, or a registration (for very small sources).

For example, in New York State, air pollution sources range in size from large industrial facilities and power plants to small commercial operations, such as dry cleaners and auto body shops. Most large sources require full air pollution permits, while smaller sources are covered by DEC's air source registration program and some mid-sized facilities require a state facility permit. For a list of major source thresholds with NY specific examples, see: *http://www.dec.ny.gov/permits/6244.html*.[10]

Any and all air emission sources should be evaluated to determine if they need to be permitted. If a facility includes a combustion facility, exhaust or process vents, or an incinerator, it may need an air permit.

The main areas that are regulated with permitting activities are

- Permits for preconstruction require that industrial sources get a permit before any construction activities begin. This allows the permitting authority (typically, the state) a chance to make sure that the appropriate emission controls are evaluated before construction begins.

- Facility operating (Title V) permits require that major industrial sources and certain other sources obtain a permit that consolidates all of the applicable requirements for the facility into one document. The purpose of Title V permits is to reduce violations of air pollution laws and improve enforcement of those laws.

- Minor source permitting or registration. When a facility does not require a Title V permit (because it is not possible for it to exceed the major source thresholds), a minor source permit or registration must be considered.

Typically, states review and issue Title V permits, minor source permits, and registrations, depending on the size and use of the facility and the discharge. EPA maintains an oversight review role for the Title V permits.

---

[10] List of Major Stationary Air Pollution Sources Under Title V at http://www.dec.ny.gov/permits/6244.html

FIGURE 1.9 Incinerator Stack. *Courtesy USEPA*

1.1.3.1 Open burning

Several states and municipalities have instituted restrictions on open burning (campfires are usually an exception). There are certain conditions where open burns can be conducted, but they are very specific, and burns must be done in accordance with the rules set forth by the regulatory authority. If anyone believes they need to conduct open burning, they should call their regional regulatory agency in charge of outdoor burning.

1.1.3.2 Waste-oil space heaters

The USEPA and most states allow the operation of a waste-oil space heater without an air resources permit, provided the facility only burns waste oil generated by the facility. The waste oil must be free of all chemical contaminants such as antifreeze, degreasers, gasoline, heavy metals, and pesticides. Waste-oil heaters must be exhausted to outside air and many states limit the size of individual heaters to 500,000 British Thermal Units (BTU) each.

FIGURE 1.10 Open Burning of Garbage. *Courtesy USEPA*

1.1.3.3 Idling of heavy-duty, on-road vehicles

Several states have passed laws making it illegal to idle heavy-duty vehicles, including trucks and buses, for more than five minutes. Generally a heavy-duty vehicle is defined as a vehicle that has a gross vehicle weight rating exceeding 8,500 pounds and is designed primarily for transporting people or property. In some states, a few exceptions exist to these rules, such as when operation of the engine is required for the purpose of maintenance, for controlling cargo temperature, or when the temperature is below 25 degrees F, and the vehicle is powered by a diesel-fueled engine.

1.2 MEDIUM (LESS PUBLICIZED) ENVIRONMENTAL PROGRAMS

1.2.1 Storm Water

On March 10, 2003, provisions of the Federal Clean Water Act went into effect, applying to many construction operations. If you are involved in construction operations that result in the disturbance of one acre or more, and storm water runoff from your site reaches surface waters (i.e., lake, stream, roadside ditch, swale, storm sewer system, etc.), the storm water runoff from your site must be covered by a either a National Pollutant Discharge Elimination System (NPDES) permit issued by the USEPA or a State Pollutant Discharge Elimination System (SPDES) permit issued by the host state.

FIGURE 1.11 Illegal Discharge to a Storm Drain. *Courtesy of USEPA*

Storm water permits are also required for operating facilities that discharge storm water, depending on the industrial activity(ies) conducted at the site.

A storm water pollution prevention plan (SWPPP) must be developed and kept on site at all times. The SWPPP is to control runoff and pollutants from a site during and after construction activities. It should specify controls to prevent erosion and sediment loading to water bodies during construction. It also should specify how, following construction, the quality of water and the quantity of storm water will be controlled.

If the proposed work involves disturbing soils at the site, the facility representative must sign a certification, notice of intent (NOI), stating that they understand and agree to comply with the terms and conditions of the SWPPP before undertaking any construction activity at the site. A copy of the NOI and a brief description of the project must be posted at the site in a prominent place for public viewing.

A NPDES permit is required to discharge storm water from construction activities which disturb one or more acres is required prior to beginning construction. If the state in which the construction is authorized to issue permits, it would be called a SPDES permit.

1.2.1.1 Exempt activities from water resources permitting

Certain activities are exempt from water resources permits in various states. For instance, in New York State, examples of exempt, non-point forestry activities are:

- Nursery operations

- Site preparation

- Reforestation and subsequent cultural treatment

- Thinning

- Prescribed burning

- Pest and fire control

- Harvesting operations

- Surface drainage or road construction and maintenance

Even if no storm water permits are needed, the facility still must adhere to all water-quality standards and prevent discharges of sediment, oil, grease, etc. into a waterway or wetland from a job site.

1.2.2 Environmental Cleanups

Under federal (and most state) laws, the identification of any contaminated area must be reported to governmental authorities when discovered.

Properties or facilities contaminated with solid and/or hazardous wastes can be cleaned up without permits because cleanups are seen as beneficial and are designed to deal with past practices. That said, all environmental cleanup activities must comply with applicable or relevant and appropriate requirements (ARARS).

For example, if the cleanup generates solid or hazardous wastes, wastewater, or air discharges, those discharges must be treated with the same

FIGURE 1.12 Hazardous Waste Clean Up Site. *Courtesy USEPA*

technologies and limitations as active sites. That means using waste hauler permits, hazardous waste manifests, wastewater treatment, and air pollution control to meet technical discharge limits.

Environmental audits of environmental cleanups should focus on the ARARS mentioned previously, to make sure no pollution results from the cleanup activities, like unsafe air releases, uncontrolled releases of storm waters and contaminated surface waters, and improper disposal of any wastes removed from the site.

A detailed discussion of the various technologies employed to clean up or contain the wastes at these sites can be found in Chapter 5 of my textbook, *Hazardous Waste Management, an Introduction.*[11]

Words of Caution for Auditors: Unknown materials and containers can be very dangerous. If handled or moved, they could cause a release of toxic materials, react with water or air, or even explode. Unknown materials could be harmful to the auditors, facility staff, or people in the area. It is best for auditors to leave unknown materials where found and request assistance to identify and deal with them. If auditors encounter unknown materials in the performance of their duties, for their personal protection, they should not handle the materials until they know what they are and how to handle them, including use of proper personal protective equipment (PPE).

Under no circumstance should an unknown material be moved until the characteristics of the material are determined. Unknown materials should not be moved, even to consolidate them with materials that seem to have the same outward appearance. If the material is a hazardous waste, it can be moved only by a properly trained individual with a permit to do so, and then only after the characteristics of the material are known. Other requirements for moving hazardous waste also apply.

Many older buildings contain hazardous materials that need special precautions and handling. Old lamp ballasts (prior to 1978) contain high levels of polychlorinated biphenyls (PCBs) and require special handling and disposal. Old window caulking materials often also contain high levels of PCBs and are a common source of contamination around older buildings,

---

[11] *Hazardous Waste Management, an Introduction,* **ISBN:** 781936420261 by Cliff VanGuilder   http://www.merclearning.com/titles/hazardous_waste_management.html

often requiring extensive cleanup operations. Asbestos can be encountered in many older buildings. Boilers and insulation installed prior to 1980 often contain asbestos. Paint manufactured prior to 1978 contains hazardous levels of lead. OHSA requirements apply in the safe and proper handling of any OSHA regulated materials. If in doubt about the safety of a situation or the contents of a container, step back and let the facility staff handle the situation. If you need to observe, do so at a safe distance.

Auditors should never rely on container labels to identify container contents, as the material in the container may not match the labeling. Hazardous waste can be placed in any container, but it may not indicate the container's proper contents or characteristics and could cause a serious health and/or safety issue.

1.2.3 Chemical/Petroleum Bulk Storage

The Chemical Bulk Storage (CBS) and Petroleum Bulk Storage (PBS) regulatory programs are similar in many ways, but have differences, which are outlined below.

The United States Environmental Protection Agency, Office of Pollution Prevention and Toxics set up industry guidance for chemical terminals and bulk storage facilities in the Emergency Planning and Community Right-to-Know Act (EPCRA) Sections 311-312.

Under the EPCRA Hazardous Chemical Storage Reporting Requirements, the federal regulations require "For any hazardous chemical used or stored in the workplace, facilities must maintain a material safety data sheet (MSDS), and submit the MSDSs (or a list of the chemicals) to their State Emergency Response Commission (SERC), Local Emergency Planning Committee (LEPC), and local fire department. Facilities must also report an annual inventory of these chemicals by March 1st of each year to their SERC, LEPC, and local fire department. The information must be made available to the public."[12]

Anyone aware of a spill should call the appropriate response center hotline. The National Response Center should be contacted at 1-800-424-8802. There is an online form that needs to be filled out as well. If the

---

[12] USEPA Emergency Management Guidance, Emergency Planning and Community Right-to-Know Act (EPCRA) Hazardous Chemical Storage Reporting Requirements at http://www.epa.gov/osweroe1/content/epcra/epcra_storage.htm

FIGURE 1.13 Leaking Containers. *Courtesy USEPA*

state in which your business is contained has a requirement for a state spill hotline, be sure to call that number as well. (If the quantity spilled is below a certain amount [reportable quantity], the spill may not need to be reported. Check with your local regulatory authority for these reporting thresholds.)

You have a responsibility to call in spills that you observe. If any employee of a company sees a spill, they should report the spill personally or notify their supervisor. Failure to report a spill of a reportable quantity of a petroleum or chemical product within the prescribed time frames can result in violations of the law, and can result in serious penalties, fines, or, in extreme cases, criminal charges, potentially resulting in incarceration.

1.2.3.1 Registration of tank and container facilities

There are different federal and local thresholds and requirements for registering tanks and containers. Following is a general list for New York State, but make sure to check with federal and local authorities to ensure compliance.

Registration is required for above-ground tanks larger than or equal to 185 gallons and all underground tanks, regardless of size. In addition, mobile or non-stationary tanks storing 2,200 pounds for more than 90 consecutive days also require registration.

Other PBS regulations include but are not limited to the following:

▪ Requirements for storing, handling, inspecting, and testing of tanks apply

▪ Additional management requirements if the tanks contain hazardous materials

1.2.3.2 Petroleum bulk storage (PBS)

The United States Environmental Protection Agency, Office of Pollution Prevention and Toxics set up industry guidance for petroleum terminals and bulk storage facilities in the Emergency Planning and Community Right-to-Know Act (EPCRA Section 313). EPCRA Section 313 Guidance for RCRA Subtitle C TSD Facilities and Solvent Recovery Facilities is contained in EPA-B-99-004 1999.[13]

This guidance requires certain industries handling and storing petroleum to report the amounts they are storing, to register certain tanks and containers, and to report the amounts of releases to the environment.

1.2.3.3 Registration of PBS facilities

There are different federal and local thresholds and requirements for registering tanks and containers. Following is a general list for New York State, but make sure to check with the federal and local authorities to ensure compliance:

▪ Registration of PBS tanks is required if greater than > 1,100 gallons.

▪ Mobile tanks (used as mobile, i.e., not permanent) are exempt from regulation.

---

[13] USEPA Publication EPA 745-B-00-002, February 2000 , Emergency Planning and Community Right-to-Know Act(EPCRA) Section 313, Industry Guidance, petroleum storage and bulk storage facilities at http://www.epa.gov/tri/reporting_materials/guidance_docs/pdf/2000/2000petro4.pdf

FIGURE 1.14 Tank Inspection. *Courtesy of USEPA*

- Heating-oil tanks having less than a combined total capacity of 1,100 gallons for on-site use are exempt from registration unless the total capacity of other petroleum-product storage tanks at the facility is greater than 1,100 gallons. In that case, even the small heating-oil tanks must be registered and compliant with the PBS program.

- All waste-oil tanks require registration (may vary by state and locality).

- Additional federal regulations for underground tanks regulations may be applicable.

Other PBS regulations include but are not limited to regular inspections, including:

- Tank and container testing

- Responsibility for transfer loadings

- Secondary containment for above-ground tanks and containers where applicable

NOTE *New York State and most other state PBS programs regulate petroleum products. The USEPA PBS program covers other types of oils as well, designated by several standard industrial codes (SIC).*

1.3 MINOR (LEAST PUBLICIZED) ENVIRONMENTAL PROGRAMS

While the remaining programs are labeled "minor" for the convenience of organization, each of the following programs are very important in a comprehensive facility environmental audit. Noncompliance with any of these programs can also cause serious environmental harm, and can lead to serious enforcement actions, including warning letters, notices of violation, monetary fines, and even criminal charges and, in extreme cases, conviction and even incarceration.

1.3.1 Pesticides

The USEPA and various states regulate the manufacture, storage, registration, and use of pesticides under the Federal Insecticide, Fungicide, and Rodenticide Act (FIFRA) passed in 1947, amended by the Federal Environmental Pesticide Control Act (FEPCA) in 1972. According to the USEPA, "The objective of FIFRA is to provide federal control of pesticide distribution, sale, and use. All pesticides used in the United States must be registered (licensed) by EPA. Registration assures that pesticides will be properly labeled and that, if used in accordance with specifications, they will not cause unreasonable harm to the environment. Use of each registered pesticide must be consistent with use directions contained on the label or labeling."[14]

In addition, most states have developed their own specific pesticide laws, rules, and regulations, at least as stringent as the USEPA standards, in order to receive federal funds to enforce the federal regulations.

---

[14] USEPA Guidance on Agriculture, Federal Insecticide, Fungicide, and Rodenticide Act (FIFRA) at http://www.epa.gov/agriculture/lfra.html

FIGURE 1.15 Farmer Applying Pesticides. *Courtesy USEPA*

Pesticides are chemicals intended to kill or control unwanted insects, animals, plants, or microorganisms. Examples of pesticides include bee and other insect sprays, weed killers, biocides for heating, ventilation, and air conditioning (HVAC) equipment, swimming pool chemicals, wood preservatives or paints with mildewcides in them, and mouse, ant, or rat poisons. Many pesticides are toxic to humans or pets. They may not be biodegradable and can accumulate in the environment.

In New York State, after January 2000, all employees applying pesticides must have proper certification as an apprentice, technician, or certified applicator. Any application of pesticides without proper certification is a violation of the regulations, unless it is an exempted activity. These certification requirements vary from state to state.

If a facility representative applies pesticides, they must follow all the laws, rules, and regulations pertaining to the application of pesticides, paying close attention to instructions on the product label.

A simple phrase unique to the Pesticides Program that sometimes helps employees and facility representatives remember the importance of pesticide compliance is:

"The label is the law!"

This phrase may seem oversimplified, but is, in fact, a direct and pertinent truth for compliance with the Pesticides Program. The product labels on all pesticides contain all instructions on the proper management, storage, and application of the product. The regulatory personnel that conduct inspections always use the label instructions as their guide to check for compliance. Any deviation from the law without written permission from the USEPA or state regulatory authority is a violation and can lead to serious consequences.

1.3.1.1 Integrated pest management

Auditors need to be aware that before a facility decides to apply pesticides to deal with pests, it should consider all other feasible alternative methods to control them. For example, if pests are there because of a moisture problem, the company protocol should be to remove all sources of moisture when possible. Also, the protocol should also ensure treatment of only the areas where pests are numerous enough to cause unacceptable damage. Auditors should also check to make sure the company minimizes the need for storage by purchasing product only in quantities necessary for treatment. Many pesticide products have shelf lives, and cannot be legally applied after their expiration date. Auditors should check the expiration dates of all pesticides to ensure they are not expired, and make sure none of the pesticides have been banned by the host state or the USEPA.

Some states have rules concerning reporting of application of pesticides by commercial pesticide applicators. In New York State, for example, all commercially certified pesticide applicators and technicians are required to complete an annual report of pesticides regardless of whether or not they applied pesticides. Prior to February 1st of each year, all pesticide use for the previous year must be submitted in writing to the supervisor. In addition to these records, commercial applicators and technicians also must maintain daily-use records of dosage rates, application methods, and target organisms for each application.

Some pesticides have been banned from use because they pose high risk to human health or the environment. It is illegal for anyone to possess or apply these pesticides. Check the label of the pesticide you plan to use against the USEPA and resident state Website, and make sure it currently is registered for the purpose you intend.

As required by the product label, all empty pesticide containers must be triple rinsed before being thrown away (or recycled), and the rinse water should be saved and used as a pesticide where practical.

1.3.2 Wetlands

Wetlands have received a great deal of attention by federal, state, and local authorities in the past several years. While there are at least four major types of wetlands (marshes, swamps, bogs, and fens), for the sake of simplicity, this text separates them into two major categories: freshwater wetlands and tidal wetlands.

NOTE

A wetland is any area that is covered or saturated by surface or groundwater often enough to support plants and other vegetation that normally live in saturated soil conditions. This means that all wetlands are not necessarily wet or covered with water. People who are trained to identify and delineate the extent and outer borders of wetlands look for those plants and determine the borders based on the existence of these plants and the borders of wet or saturated areas.

1.3.2.1 Freshwater wetlands

If a facility plans to do work in or around (within 100 feet) an area that is wet during part or all of the year, or that has plants typically found in wet places, the area may be a wetland regulated under the U.S. Corps of Engineers, or any state or other local governmental authority.

Work in these areas typically requires permits from one or many of these regulatory bodies. The wetlands are typically delineated on maps at the federal and state level, and in many cases, these maps do not necessarily

Where water meets land

FIGURE 1.16 Freshwater Wetland. *Courtesy USEPA*

agree, with the state or local authorities generally identifying more wetland areas than the federal maps.

In New York State, for instance, wetlands regulated by New York State (greater than 10 hectares in area) are shown on final freshwater maps for each county. Some wetlands may be regulated under New York State Law Article 15. These wetlands are located along watercourses or lakes, with permit review done by New York State Department of Environmental Conservation fisheries staff.

If you suspect a regulated wetland is in the vicinity of a work area, it is highly advisable to contact either the federal or local regulatory staff to request a wetland determination or delineation. In many states, this delineation can be done by individuals trained to delineate wetlands, but a final review by regulatory staff is almost always required. After a request by an applicant, regulatory staff will decide whether a state regulated wetland exists on the property and point out the boundary (determination) or flag the boundary (delineation).

While only certain wetlands are regulated by state and local governments, the U.S. Army Corps of Engineers (USACOE) regulates certain activities in all wetlands under Section 404 of the federal Clean Waters Act. If the facility plans to work in or around a wetland, it is imperative that USACOE make a determination of jurisdiction. It is highly inadvisable to attempt to make a wetlands determination without regulatory confirmation, as the work done without a permit may be in violation, subjecting the facility and owners to serious fines and requiring remedial action.

1.3.2.2 Tidal wetlands

Tidal wetlands are separated from freshwater wetlands because they are in areas affected by ocean tides. This means they often contain some saltwater, and are regulated differently from freshwater wetlands under federal and some state and local laws as well. Typically, a facility planning certain activities within or around these wetlands will require a permit.

Examples of activities requiring a permit in these areas include any kind of clearing, draining, discharging, grading, filling, or construction. Facilities may also require freshwater permits as well, as the area in question may include freshwater wetlands. Other permits from the USACOE may be required in wetlands adjacent to navigable waters or protected watercourses.

FIGURE 1.17 Tidal Wetland. *Courtesy USEPA*

Jurisdictional determinations under federal wetlands statutes must be made by USACOE for all wetlands, including tidal wetlands.

1.3.2.3 Federal- or state-protected wetlands

State wetlands are designated on many state wetland maps, and generally have a lower size limit than the federal limit. For example, in New York State the regulatory threshold for wetlands is 12.4 acres.

Federally protected wetlands maps are also generally available, but are more complicated to identify than state-protected wetlands, and there is no lower size limit. Permits for state wetlands (if applicable) would be issued by that authority. The USACOE issues permits for all federal wetlands.

FIGURE 1.18 Federally Protected Wetland. *Courtesy USEPA*

1.3.3 Stream Disturbance

A permit is needed to work on the bed or banks of a protected stream. Permits are not required for work in a non-protected stream, but water quality standards like turbidity and sedimentation must be maintained. Auditors should make sure the company consults with local regulatory agencies before working on a stream.

1.3.4 Lands and Forests

Before trimming brush or removing trees or shrubs from a property, auditors should make sure the company reviews all local laws, rules, and regulations concerning these activities. It is also wise to check with a local forester or forestry technician to verify the company's understanding of permissible activities. Prior to any tree or brush cutting, it may be necessary to know the land classification, the boundary lines of the property, protected plants in the area, and any special rules that may be applicable.

FIGURE 1.19 Disturbed Stream Bank. *Courtesy USEPA*

1.3.4.1 Property boundaries

Before performing any construction work or ground disturbance, the facility should make sure all property boundaries are clearly marked, and all contractors and staff know all property boundaries. If the facility has not had an official property boundary survey, it is wise to have one completed to make sure the existing facilities are all within the property boundaries and have the proper legal setbacks. Hazardous waste storage facilities must be at least 50 feet from the facility property line.

While property boundary issues may apply to any physical activity, such as construction, or cleanup, it was listed here because forestry activities are generally the most common to result in trespassing. This can be a serious liability issue with potentially costly consequences.

1.3.4.2 Protected plants

Before construction or ground disturbance, the company should identify all rare, threatened and endangered plants, and plants vulnerable to exploitation. A picture of a protected plant (lady slipper) is shown in Figure 1.20.

This may require the help of a local botanist or forester. Often, staff from a local regulatory agency will assist in locating and identifying any protected plants.

FIGURE 1.20 Lady Slipper. *Courtesy USEPA*

1.3.4.3 Stream crossings

If a facility needs to build any structure to cross a stream, a stream crossing permit may be required. If needed, this permit application would be made, reviewed, and issued by the USACOE or a state or local authority, depending on the regulations in that area.

1.3.5 Mining (Mined Land Reclamation)

Regulation of the mining sector is handled by individual states. While the USEPA does not regulate mining activities, mining often involves every major USEPA regulatory program.

Mining operations generate wastewater discharges, air emissions, tailings, and waste rock that must be disposed. As a result, discharge permits may be required to protect surface and groundwater quality, drinking water, and air quality. Active and abandoned mines can also cause extensive losses of natural habitat, so facilities should evaluate those impacts and mitigate the effects. The federal government started regulating mining activities as early as 1891, followed by an extended evolution of federal legislation. Ironically, the federal government leaves the permitting of mining activities to individual states.

FIGURE 1.21 Mining Operation. *Courtesy of USEPA*

Most states have actively regulated mining activities since the 1970s, and states with coal mines have been regulating coal mining as early as 1910. State mining reclamation programs are set up to ensure sound environmental management principles and that land is returned to productive use after mining.

In most states, each mining applicant must submit a mining permit application and a land use plan. The application is a standard form, and the mined-land use plan generally consists of two parts:

1. Mining plan, which explains the proposed mining method

2. Reclamation plan, which explains the proposed method of reclaiming the land

There are normally quantity thresholds that trigger permit requirements. For instance, in New York State, any person who mines or proposes to mine more than 1,000 tons or 750 cubic yards of minerals, including dirt or gravel, within 12 successive calendar months must apply for a permit. Certain activities may be exempt from the permitting requirements of the Mined Land Reclamation Law. A potential applicant should contact a regulatory specialist to determine whether an activity requires a permit. This permit is an approval to conduct regulated activities at a specific site.

1.3.6 Wildlife

1.3.6.1 Protected species

The federal government identifies certain species of animals and fish as protected or endangered under the Endangered Species Act. If a company plans to construct or expand any part of its facility, it is prudent to check with local regulators about whether the proposed construction might affect any wildlife protected under federal, state, or local laws, rules, or regulations.

1.4 REGULATORY DIFFERENCES FROM STATE TO STATE

As mentioned previously, the federal government has passed laws and developed policies, rules, and regulations to deal with every environmental program possible. In almost every one of these programs, the federal government has set up detailed delegation plans to delegate part or all of these to individual states. This federal delegation comes with federal funds to the states if they can prove they have adequate resources to implement these rules, and pass laws and develop rules and regulations that are at least as stringent as the federal laws, rules, and regulations.

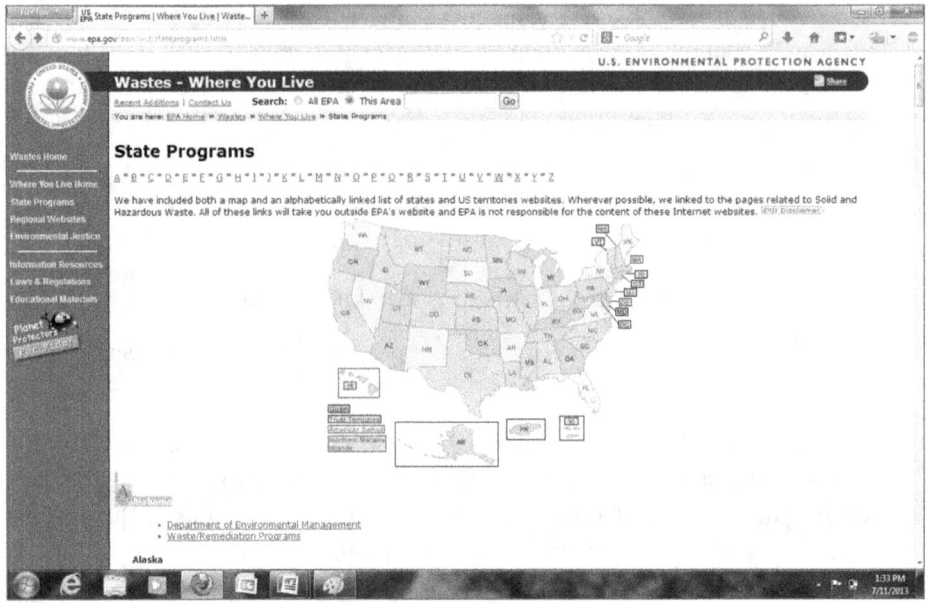

FIGURE 1.22 EPA Website Portal for State Programs. *Courtesy USEPA*
http://www.epa.gov/osw/wyl/stateprograms.htm

This has led to a very complicated regulatory environment for businesses trying to comply with all the rules, especially when those businesses have facilities in more than one state (or country).

It is obviously imperative that the facility comply with all federal laws, rules, and regulations, but equally important that they understand and comply with all state and local laws, rules, and regulations as well.

An illustration of USEPA's Website page for finding specific state regulations is illustrated in Figure 1.22.

1.5 RELATED PROGRAMS

Up to this point in the chapter, the text has been describing the regulatory requirements set up for the protection of the environment by controlling pollution sources. There are several programs that are voluntary in nature which provide environmental protection, and should be considered in every comprehensive environmental audit.

1.5.1 Sustainability

A great deal has been written and discussed about a popular new term of art called sustainability. There are numerous definitions of sustainability, but leaning toward simplicity, we have defined it as "developing and implementing programs designed to sustain a safe, clean environment and lifestyle for all future generations." Due to the subjective nature of this term, the federal government has not developed any laws, rules, or regulations to directly affect this goal. It is arguable that the overall federal regulatory rubric sets up parameters that help preserve a safer environment.

The more subjective and difficult issue is the lifestyle or quality of life issue. Scientists and social engineers argue constantly about the effects of human activities on our environment, including depletion of natural resources and changes to the earth's climate. It is not the purpose of this text to agree or disagree with the merits of these arguments. The purpose of this text is to recognize the potential negative effects of failure to follow environmental laws, rules, and regulations.

That said, it should be the goal of all facilities to meet and exceed environmental standards, with the goal of controlling pollution to a point where the negative effects on human health and the environment are minimized, or even reversed. This takes a serious commitment from management and staff, but it is possible. The next section discusses ways this can be accomplished.

1.5.2 Waste Management Hierarchy

In 1989, after observing the trend of land burial of solid wastes and hazardous wastes without adequate treatment, the USEPA set up a waste management hierarchy. This hierarchy is illustrated below and described in more detail in the following sections.

1.5.2.1 Source reduction and reuse

These two terms are at the top of every waste management hierarchy.

Source reduction can be defined as "reducing waste at the source of production or even before production starts." Reuse can be defined as reusing by-products or waste products in the same manufacturing process. At the very beginning of a manufacturing process, every facility owner/manager is faced with several major decisions before starting to make products, including, but not limited to:

- What product(s) do I want to make?

- How many of these products should I make?

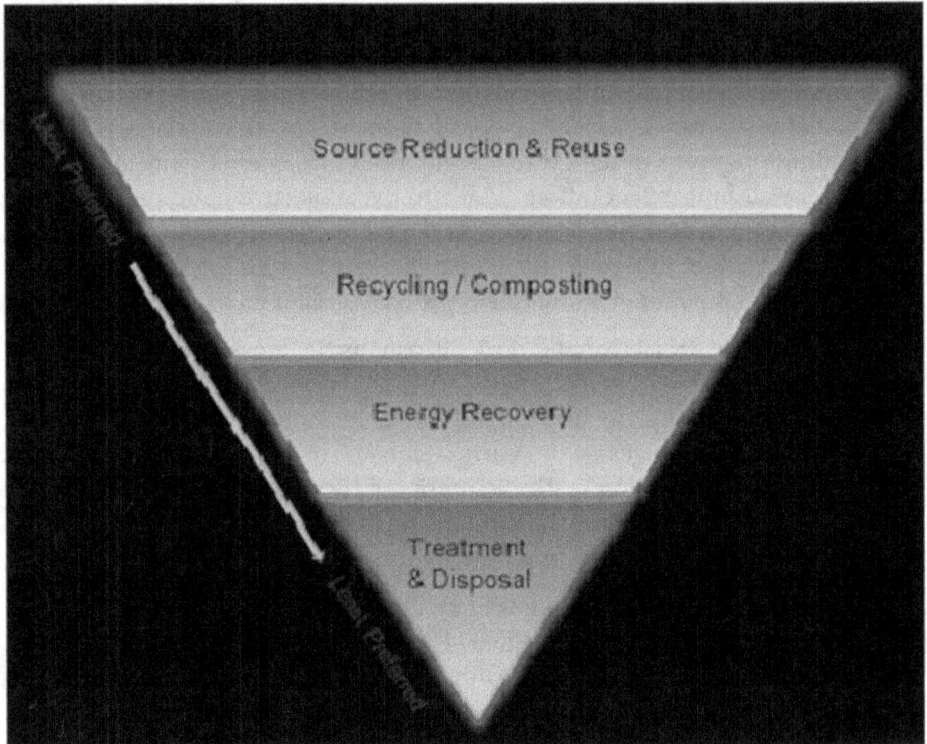

FIGURE 1.23 EPA Waste Management Hierarchy. *Courtesy USEPA*

- What ingredients do I need?
- What are the least toxic ingredients I can use?
- What recipe produces the least amount of toxic waste?
- What equipment do I need?
- How many people do I need and what training do they need?
- What waste streams will I generate and how do I manage them?

In each of those decisions, the facility representative can help protect the environment and encourage sustainability. For instance, if the safest, least-toxic ingredients are used, the by-products and waste should be less toxic. Likewise, if the minimum amount of products are ordered with little surplus, the amount of ingredients, by-products, energy, and waste are reduced commensurately. If the equipment is energy efficient and properly designed and built for its intended purpose, more savings result. If the personnel responsible for making the products are properly trained, they will make the products more efficiently, and may contribute ideas to streamline the process or produce less by-products and waste. Finally, an analysis of the waste streams should be accomplished, with an evaluation of how to reuse or recycle part or all of the waste. If this is not possible, ways should be evaluated to make the waste less toxic or more amenable to treatment.

1.5.2.2 Waste recycling

After a facility has fully exhausted all source reduction options, it should evaluate ways to recycle the by-products (intermediates) and the wastes. Recycling is the process of using waste products or by-products as feedstock to the same process or another process to make another useful product. Sometimes the waste from one process is of direct use by another consumer, which excludes it from ever being a waste by regulation. (Note that the USEPA waste management hierarchy consider energy recovery and treatment to be equal priorities.)

1.5.2.3 Energy recovery

This third tier of the waste management hierarchy is to encourage facilities to use the energy available in wastes that cannot be reduced, reused, or recycled to heat processes, facilities, or any other areas that require heat rather than bury the organic portion in a landfill. Examples include:

- Waste to energy facilities

- Lightweight aggregate facilities

- Composting facilities

1.5.2.4 Treatment and disposal

The fourth and final stage in the waste management hierarchy is the treatment and disposal of any residue that cannot be reduced, reused, recycled, or used for energy recovery.

There are too many forms of treatment to be addressed in this text, but the auditor(s) should be familiar with the treatment processes. The auditor should consult with design and operation professionals to ensure that the waste management hierarchy is being utilized and that the treatment is being done in accordance with all of the laws, rules and regulations, and if possible, that the best available technologies are being employed.

A detailed description of available waste treatment and disposal technologies is available in my other textbook, *Hazardous Waste Management, an Introduction.*[15]

NOTE

The federal Hazardous Waste Land Disposal Restriction (LDR) Program distinguishes between hazardous waste treatment and hazardous waste disposal, citing that wastes should be treated to the extent possible before land disposal of only treated residuals. The federal solid waste policy does not make that distinction.

1.5.3 Leadership in Energy and Environmental Design (LEED) for Buildings

While it is not a regulatory program, it is prudent for all comprehensive environmental audits to address the LEED (Leadership in Energy and Environmental Design) program.

LEED is an internationally recognized method of third-party verification that a building is being built in an environmentally friendly and energy efficient (green) manner. It is a voluntary, consensus-based, market-driven program that is helping transform the design, construction, and operation of buildings to a more environmentally friendly, energy efficient manner. LEED is comprehensive and flexible and addresses the entire lifecycle of a building. Because it entirely voluntary, auditors need not be concerned with the liability associated with environmental violations arising from this program. However, if there are environmental benefits to be derived from implementing recommendations from this program, the auditors should include them in their comprehensive report wherever possible.

1.5.4 Homeowner Environmental Audits

While this book is written primarily for industrial facilities, a section has been added on environmental audits for homes. While this may seem unnecessary, homeowners are exempted or excluded from almost all environmental laws, rules, and regulations, and by an ironic twist from

---

[15] *Hazardous Waste Management, an Introduction, ISBN: 781936420261* by Cliff VanGuilder http://www.merclearning.com/titles/hazardous_waste_management.html

FIGURE 1.24 Building with LEED Features. *Courtesy USEPA*

governmental generosity, are generally unaware of the hazards of the toxins and the attendant health effects to which they are exposed every day.

For example, no federal agency, including the United States Environmental Protection Agency and the United States Food and Drug Administration (USFDA) regulates household cleaners. This means household cleaners and other chemicals used in homes can be highly toxic, requiring the homeowner to use personal protective equipment and special safety precautions that may not be available to them and for which they have not been properly trained. The personal protective precautions given on the homeowner product label are sometimes far less protective than for the exact same chemical when handled in an industrial setting.

An environmental audit of a home can produce startling results, alerting the homeowner to everyday household cleaners that are acidic, caustic, reactive, or flammable, and making them aware of the possible dangers associated with handling these chemicals.

Provided is a comprehensive list of these household toxins as a bonus to this text in the author's e-book entitled *Detox Your Home—Banish the Toxic Menace.*[16]

Lists in this e-book are convenient references for conducting an environmental audit of a home or a standard office with no industrial activities.

This bonus e-book also describes the health hazards associated with each household toxin, and provides alternative cleaners that are much less toxic or completely safe for homeowners to use.

Added features of this e-book include chapters on naturally occurring toxins that may be present in homes, such as radon gas, and toxins that may already be present in homes that are purchased, like lead-based paint, asbestos, and other toxins.

Home environmental audits are strongly recommended to educate the homeowners and nonindustrial offices on the toxins present, safer alternatives that may be employed, and toxins that may already be present in these buildings.

In Chapter 2, this text will describe how to determine the necessary and proper scope of work needed for an environmental audit.

---

[16] *Taming the Toxic Menace in Your Home*, copyright registration # TXu 1-671-612 at http://www.amazon.com/Taming-Toxic-Menace-Your-ebook/dp/B008G65RWQ

2

DETERMINING NECESSARY AND PROPER SCOPE OF WORK

2.1 BUSINESS PLAN

In order for any business to survive economically, it needs a business plan (written or unwritten), and this business plan needs to be followed or the business faces eventual financial failure. This is the business's plans for survival and success. The plan includes products or services, asset management, income and expenses, etc. Because the subject of this book is environmental audits, this text will not address all aspects of business planning. Instead, it will focus on environmental asset management, and the underlying expenses associated with environmental audits.

2.1.1 Asset Management

The assets of a company are essential to any business plan, and the company's ultimate financial success or survival. There are many asset management computer programs that can be found and purchased either online or at any stores where computer software is sold. Because asset management is not included in the scope of this text, this book does not recommend any particular asset management program. However, the author recommends all businesses explore the concept of asset management and carry out at least a rudimentary asset management plan and audit.

When an asset management plan is carried out, the company will identify, included within these assets and expenses, wastes produced and the cost of managing and disposing of those wastes. They will also find all waste management equipment, from waste storage facilities to wastewater treatment facilities, to incinerators, etc.

Whether wastes are seen as an asset or a liability, they must be recognized as an essential by-product of the business process, so the waste processing equipment and all waste management costs must be included in the company's business plan.

2.2 SCOPE OF ENVIRONMENTAL AUDIT(S)

If a company independently decides an environmental audit is prudent or necessary, the scope of the audit must first be determined. A list of possible scopes is listed, along with attendant descriptions of the scopes and the possible reasons for an audit. As illustrated, in some cases, the audit is voluntary, and in others it may be mandatory.

2.2.1 Comprehensive Audit

The only reliable method for a facility to ensure total compliance with all environmental laws, rules, and regulations is to conduct a comprehensive, facility-wide environmental audit. Depending on the size and complexity of the facility, this may be an involved, lengthy, and potentially expensive process. However, as mentioned previously, environmental compliance costs represent required business expenses that must be included and factored into any business plan to accurately represent operating costs. Failure to recognize these costs and include them in a business plan can have disastrous results, up to and including financial failure of the business.

A comprehensive environmental audit is designed to identify and evaluate

- All wastes generated by the facility (air, water, and solid and hazardous waste).

- The processes that generate those wastes.

- The methods used to manage those wastes.

- Whether the facility is managing those wastes in the most efficient and cost-effective manner.

- Whether the facility complies with all applicable waste management laws, rules, policies, and regulations.

As illustrated in Chapter 1, environmental compliance extends far beyond rudimentary waste management for the protection of air, water, and land. It also includes areas like chemical and petroleum bulk storage, spill prevention and reporting, pesticide use, wetland protection, stream disturbance, etc.

Any comprehensive audit should evaluate all waste generation and management for adherence to the waste management hierarchy; that is source reduction and waste minimization first, followed by reuse and recycling second, treatment of all wastes third, followed by the disposal of only treated residuals. This waste management hierarchy is a good tool to help ensure that the wastes that are generated represent the minimum quantities and lowest toxicity possible.

Another valuable voluntary program that can help a facility with environmental compliance is the LEED program. Described in Chapter 1, LEED is a method of third-party verification that a building is being built in an environmentally friendly and energy efficient (green) manner. It can help ensure the design, construction, and operation of buildings to a more environmentally friendly, energy efficient manner.

For a large facility, a comprehensive environmental audit can be a complicated, lengthy, and potentially expensive process. However the benefits of understanding all of the environmental aspects of the facility can far outweigh the costs, particularly if the facility makes sure all of the recommendations of the audit are implemented. If the audit is truly comprehensive, and is done in a thorough and accurate manner, it provides the facility owner/operator with the knowledge that the facility is causing the minimum environmental impacts possible, and that any regulatory compliance audits will be successful, with minimum or no violations.

2.2.2 Required Scope from Enforcement Action

If an audit is required by an enforcement action resulting from a regulatory inspection, the minimum scope will be predetermined by the enforcement action. This scope will obviously be prescribed to meet the requirements of the consent order, consent decree, court order, etc.

However, that does not necessarily limit the scope of the audit. The facility may elect to conduct a more extensive audit. This decision should be made with the caution that if the scope is broadened beyond that of the requirements of the enforcement action, the variations or noncompliance findings could cause complications in the enforcement action. This can be avoided if the enforcement staff agrees in advance to allow findings of noncompliance from the broader scope to be corrected without additional penalties.

Many audits prescribed by enforcement actions require the facility to have the audit conducted by an independent auditor, often a licensed professional engineer, to ensure that no conflicts of interests occur and that the results of the audit are totally objective. The facility often is given the right to review and comment on the audit findings before the final report is turned over to the regulatory agency. Depending on the enforcement agreement, the facility may be allowed to correct any factual or technical discrepancies to the enforcement agency when the audit report is submitted. Often, the facility is required to report those discrepancies when they submit the audit report.

If the enforcement agreement allows the facility to review the audit report before submittal to the enforcement agency, they should conduct a careful and thorough review of the entire audit report, and make sure it is factually accurate. The author was personally involved in reviewing several audit reports required by enforcement actions, and in more than one case, the facility did not catch and correct incorrect information about the waste management practices at the facility, and that incorrect information caused further review and questions.

2.2.3 Audits of Select Environmental Programs or Parts of Programs

A facility can decide to conduct program-specific environmental audits at any time for a number of reasons:

- Pre-emptive inspection audits. Some regulatory enforcement inspections are conducted on a regular basis (annual, biennial, quarterly, etc.). If the timing of a regulatory inspection is predictable, it is prudent to conduct a pre-inspection audit to ensure substantial compliance before the regulator(s) arrive(s).

- Internal routine audits. It is prudent for facilities to have a schedule for conducting audits on a routine basis, even if the audit is a brief checklist.

- Updated regulations audit. After a regulatory change, facility procedures may change for one or more staff. Conducting these update audits keeps staff aware of the latest regulatory requirements and helps ensure the activities are in compliance.

- Voluntary compliance audits. Staff will sometimes suggest new or revised procedures to improve waste management or environmental compliance. If these changes are implemented, an audit can be conducted to make sure the procedures are in place and give the staff recognition for their idea(s).

2.2.3.1 Cost of environmental audits

Cost is a critical factor in deciding what programs to audit and at what intervals. The costs may be spending in-house staff time, hiring outside consultant costs, or a combination.

Generally, if the audits are done by in-house staff, the facility can save substantial consulting costs by contacting environmental regulators with questions on compliance. In the author's experience, if the facility hires outside consultants to conduct environmental audits, the consultants contact regulatory staff on a regular basis, and often do so without divulging the identity of their clients. In-house staff can ask the same questions anonymously, but there is sometimes an inherent fear or mistrust of government officials discovering the facility's name and pursuing the potential violations. This was not the case in the author's personal experience. Environmental regulatory staff is generally more than willing to help any facility understand the rules and help them comply.

The USEPA provides extensive, detailed guidance on environmental audits online, including:

- Environmental Protection Agency's Incentives for Self-Policing: Discovery, Disclosure, Correction, and Prevention of Violations; Notice[17]

- Guidance prepared by the Federal Facilities Enforcement Office (FFEO), U.S. Environmental Protection Agency (USEPA) for federal facilities to use as guidance in establishing and implementing environmental audit programs[18]

- Compliance Incentives and Auditing, Audit Protocols for several environmental programs, including:

  - EPA New England Inspectors' Multimedia Checklist[19]

---

[17] Environmental Protection Agency's Incentives for Self-Policing: Discovery, Disclosure, Correction and Prevention of Violations; Notice Federal Register, Vol. 65, No. 70, Tuesday, April 11, 2000

[18] Guidance prepared by the Federal Facilities Enforcement Office (FFEO), U.S. Environmental Protection Agency (EPA) for federal facilities to use as guidance in establishing and implementing environmental audit programs at http://www.epa.gov/compliance/resources/policies/incentives/auditing/envaudproguidemas.pdf

[19] EPA NEW ENGLAND INSPECTORS' MULTIMEDIA CHECKLIST at http://www.epa.gov/region2/capp/cip/r1mm99.pdf

---

EPA NEW ENGLAND INSPECTORS' MULTIMEDIA CHECKLIST

General Information

Inspector: _____   Date: _____

Facility Name: _____   Contact: _____

Address: _____
         (STREET)   (CITY)   (STATE)   (ZIP)

Phone No.: (___)___-_____ SIC Code: _____ No. Employees: _____

Products mfd. and description of facility: _____
_____

Air: Stationary Source Compliance

1.O   Did you observe opaque smoke emitted from a smokestack (dark enough to obscure anything behind the plume)? _____ - If yes - Which process line (be specific, i.e., boiler No. 4)? _____ - Air pollution control equipment out of service? _____ - If yes - When will it be back on line? _____

2.OP   Did you smell any strong odors? _____ If yes, from what process? _____ What chemicals (i.e., solvents) were causing the odors? _____ Is the process controlled by air pollution control equipment? _____

3.I   Has the facility added any processes which emit air pollutants or expanded any pre-existing air-pollution emitting processes in the last two years? _____ - If yes, what type of process was added? _____ Did the facility obtain a state air permit? _____

4.I   Does the facility operate a degreaser(s) that uses one of the following cleaning solvents:
        Methylenechloride   Perchloroethylene   Trichloroethylene
        111-Trichloroethane   Carbontetrachloride   Chloroform
    If yes, has the facility submitted an initial notification and a notification of compliance to the EPA? _____

EPCRA N313 (Spill Notification/Chemical Inventory) and **CAA 112(r)** Risk Management Plans

1.I   Has the facility experienced any accidental releases above the Reportable Quantity (RQ) within the last three years? _____ If yes, provide the name of the chemical released, the quantity and the date. _____

2.I   Does the facility have on-site at any time during the calendar year a) 10,000 lbs or more of any hazardous chemical requiring an MSDS or b) a threshold reporting quantity of a listed Extremely Hazardous Substances (EHS)? _____

3.I   If yes, have Tier II chemical inventory forms been filed annually with the fire department and local/state planning authorities? _____

4.I   Does the facility have more than a threshold quantity of a listed chemical in any individual process (including storage) requiring preparation of a Risk Management Plan?

5.1   If yes, has the facility submitted an RMP summary to EPA via the Internet?

FIGURE 2.1 EPA New England Inspector's Multimedia Checklist. *Courtesy USEPA*   *(Continued)*

EPCRA Section 313 (Toxic Release Inventory)

1.I   Does the facility manufacture, process, or use any toxic chemicals in a quantity greater than 10,000 lbs per year?

2.I   Has the facility submitted any toxic chemical release forms (Form R) to EPA?

FIFRA

1.I   Does the facility manufacture, distribute, repackage, relabel, store or use pesticides? (Product which would be considered pesticides includes disinfectants, sterilizers, germicides, algicides, virucides, swimming pool compounds, insecticides, fungicides, herbicides, etc.) _____

2.I   If yes, does the label bear an EPA registration and establishment number? _____

RCRA

1.I   Does the facility generate or otherwise handle hazardous waste? If so, describe the types of hazardous waste generated/handled, and state whether it is generated on-site or received from off-site. _____
_____

2.OP   Do you see any waste stored in containers, drums, tanks, pails, or dumpsters? Note the approximate quantity of waste, and its location. _____
_____

3.OP   Are there any containers or tanks of hazardous waste that are open or in poor condition(leaking, corroded, etc.)? If so, describe waste (e.g., liquid, sludge, etc.), indicate markings on containers/tanks and the container/tank location(s). _____
_____

4.OP   Is there any evidence of spills or leaks or dumping to the ground, pits or lagoons? If so, note location and extent of release. _____
_____

5.I   Does the facility operate a boiler or industrial furnace? Has there been any incineration of hazardous waste on-site? If so, what type of hazardous waste, and is this an ongoing operation? _____

SPCC

1.I   How many gallons of oil does the facility store above and below ground? _____
_____

- If the facility stores more than 660 gallons in a single tank or more than 1320 gallons in a number of tanks above ground or more than 42,000 gallons below ground - Does the facility have a certified SPCC (Spill Prevention, Control, and Countermeasure) plan signed by a P.E.? _____ Date of plan: _____. Certifying P.E. No. _____.

FIGURE 2.1 EPA New England Inspector's Multimedia Checklist. *Courtesy USEPA*   *(Continued)*

TSCA PCB

1.OI   Is there any evidence of liquid-filled electrical equipment that may contain PCBs?

If "yes," describe type of equipment: _____

_____

_____

2.IP   If the above equipment is considered to contain PCBs, what was the basis for this determination:
based on- marking with "Large PCB Mark"? Yes: _____ No: _____
based on- equipment Nameplate? Yes: _____ No: _____
based on- information from facility rep.? Yes: _____ No: _____

3.OP   Is there any evidence of spills or leaks from transformers, capacitors, or other liquid-filled electrical equipment that may contain PCBs? Yes: _____ No: _____
If "yes," describe type of equipment and spill or leak _____

_____

4.OP   Are there any PCB items (equipment, drums of waste or other containers) in storage for disposal? Yes: _____ No: _____
Where are these items being stored, and what is their condition? _____

_____

TSCA Core

1.I   Does the facility manufacture (synthesize anew) any chemical substances in any amount?
_____ If so, in simple terms, what chemical(s) do they make? _____

2.I   Does the facility import any chemical substances into the United States? (Company is "Importer of Record") _____

UST

1.I   Does the facility store in USTs motor fuels, waste oils, and/or hazardous substances?
_____ YES _____ NO (Note: USTs containing heating fuels for on-site heating purposes are exempted from RCRA UST.)

If Yes, ask:

2.I   Are the USTs registered with the state? _____ YES _____ NO (Each state keeps notification data for USTs)

3.I   Is some form of leak detection in use for the UST system's tank and associated piping)?
_____ YES _____ NO

4.I   Are records available showing registration and monthly leak detection along with the yearly UST system tightness test? _____ YES _____ NO

FIGURE 2.1 EPA New England Inspector's Multimedia Checklist. *Courtesy USEPA*   *(Continued)*

Water

A. DIRECT (NPDES) & INDIRECT (PRETREATMENT) DISCHARGERS

1. I   Has the facility expanded its production or wastewater flow, or changed its processes, since its last permit? _____

O   Did you observe any outfalls? _____ If so, was there any discoloration, steam, oil sheen, or odor? _____

2. P   Does the facility use water in its manufacturing processes? _____
If yes -

I a   Does the facility discharge wastewater (process, sanitary, cooling, etc.) into a surface water, municipal sewer system, or a subsurface system? _____ Is the municipal sewer separate or combined? _____ .

I b   Are all of the discharges covered by a permit? _____

3. P   Does the facility have floor drains? _____ If yes -
a) Are materials stored in a manner that leaks or spills could enter the floor drains? _____

b) Are materials dumped down the floor drains? _____

I   c) Where do the floor drains discharge (1. treatment facility, 2. municipal sewer, 3. subsurface system, 4. storm drain or 5. surface water)? _____

4.   Does the facility treat its process and/or sanitary wastewater prior to discharge? _____
If yes -

OP   a) Is the treatment equipment operational, clean, and well maintained? _____

OP   b) Is the discharge free of solids, color, and odor? _____

B. STORM WATER

1. O   Are there catch basins, drains, culverts, ditches, etc. on the property intended to convey storm water. _____ If yes - Is the storm water conveyed to a (1) treatment facility, (2) combined sewer, (3) separate storm sewer, (4) separate sanitary sewer, or (5) surface water? _____

2. I   Are the storm water discharges covered by a permit or has the discharger applied for a permit? _____

3. O   Are materials stored outside? _____ If yes -
a) Are materials (1) stored in sealed containers, under tarps or roofs, or (2) are they open to contact with precipitation? _____
b) Are outside material handling/storage areas clean and kept in a manner to prevent contamination of runoff? _____ .

Wetlands

1.O   Within view, are there a) streams, ponds or other water bodies; b) vegetated areas with standing water; or c) areas with mucky, peaty, or saturated (squishy) soils? If yes, have any of these areas been disturbed by waste/refuse disposal, storage of materials, ditching, or filling? If yes, briefly describe:

_____

FIGURE 2.1 EPA New England Inspector's Multimedia Checklist. *Courtesy USEPA*   *(Continued)*

2.I   If yes to both "Observable" questions, does facility have a federal CWA section 404 permit, or any state or local permit authorizing the activity(ies) observed?

<hr/>

Underground Injection Control (UIC)

Subsurface Wastewater Disposal

1.A. [I] Does the facility discharge fluid wastes to drains, plumbing, or drainage systems connected to a subsurface wastewater disposal system(s) listed below? Yes: _____ No: _____
If "yes," circle discharged fluid waste type(s) in list below.
If "yes," circle subsurface wastewater disposal system type(s) in list below.

1.B. [I] Does the facility have a state or local permit authorizing the subsurface wastewater disposal system(s)? Yes: _____ No _____

Subsurface Sewage Disposal

2.A. [I] Does the facility have an onsite sewage disposal system(s) that serves more than 20 people per day? Yes: _____ No _____

2.B. [I] Does the facility have a state or local permit authorizing the onsite sewage disposal system(s)? Yes: _____ No _____

2.C. [I] Is the onsite sewage disposal system used to dispose of any fluid waste type(s) listed below? Yes: _____ No _____
If "yes," circle discharged fluid waste type(s) in list below.

3. [O] Do you observe any floor drain(s) or surface drain(s) that can receive any of the fluid waste type(s) listed below? Yes: _____ No _____
If "yes," check the discharged fluid waste type(s) in list below.

3. [I] If "yes", where does the facility drain(s) discharge: a. Municipal sewer? _____
b. Subsurface disposal system type(s) listed below? _____ c. Unknown _____

Disposal System Type

dry well	septic system	cesspool	leach field	leach pit	leaching trench	disposal well

Fluid Waste Type

Liquid Wastes	**Wash Water, Spills or Storm Water from:**
Process Wastewater	* maintenance areas
Non-Contact Cooling Water	* hazardous material or waste storage areas
Boiler Blowdown Fluids	* hazardous material or waste handling areas
Air Conditioning Fluids	* process or manufacturing areas
Heat Pump Fluids	* fuel storage areas
Other Waste _____	* areas prone to hazardous substance release

FIGURE 2.1 EPA New England Inspector's Multimedia Checklist. *Courtesy USEPA*   *(Continued)*

CODES:

O - OBSERVABLE   P - PROCESS   I - INTERVIEW
Pollution Prevention (Optional)

1.I   Does your facility have a pollution prevention, toxics use reduction, or RCRA waste minimization program?

Would you be interested in finding out more about pollution prevention techniques?
Take a look at this brochure, and feel free to call any of the EPA, state, or other staff listed for more information.

ADDITIONAL COMMENTS

_____

_____

CONCLUSION

_____

_____

RECOMMENDATION FOR FOLLOW-UP

_____

_____

Forwarded to Contact		Phone	Mall Code
Air-stationary	Fred Weeks	918-1855	SEA
EPCRA 313	Cynthia L. Brown	918-1743	SEA
TSCA Core	Rose Toscano	918-1861	SEA
EPCRA (Non313)	Don Mackie	918-1749	SEA
FIFRA	Wayne Toland	918-1852	SEA
RCRA	Ken Rota	918-1751	SER
SPCC	Don Grant	918-1768	SEW
TSCA PCB	Abdi Mohamoud	918-1858	SEA
UST	Bill Torrey	918-1311	HBO
WATER	Joan Serra	918-1881	SEW
UIC	Dave Delaney	918-1614	CMU
WETLANDS	Denise Leonard	918-1719	SEE
RMP	Ray DiNardo	918-1804	SPP

FIGURE 2.1 EPA New England Inspector's Multimedia Checklist. *Courtesy USEPA*   *(Continued)*

ACRONYMS

AIR
AFS - AIRS Facility Subsystem (EPA's air compliance database)
AIRS - Aerometric Information Retrieval System
BACT - Best Available Control Technology
CAA - Clean Air Act
CAAA - Clean Air Act Amendments
CEM/CEMS - Continuous Emission Monitoring/System
CFC - Chlorofluorocarbon
EER - Excess Emission Report
HAP - Hazardous Air Pollutant
HON - Hazardous Organic NESHAP
LAER - Lowest Achievable Emission Rate
NAAQS - National Ambient Air Quality Standards
NARS - National Asbestos-Contractor Registry System
NESHAPS - National Emission Standards for Hazardous Air Pollutants
NSPS - New Source Performance Standards
NSR - New Source (Pre-construction) Review
PM - Particulate Matter
RACT - Reasonably Available Control Technology
SIP - State Implementation Plan
VE - Visible Emissions
VOC - Volatile Organic Compounds

EPCRA
EPCRA - Emergency Planning and Community Right-to-Know Act
LEPC - Local Emergency Planning Committee
SERC - State Emergency Response Commission
TRI - Toxic Release Inventory

FIFRA
FIFRA - Federal Insecticide, Fungicide, and Rodenticide Act, as Amended
EPA Reg. No. - EPA Registration Number (one for each pesticide)
EPA Est. No. - EPA Establishment Number (where a pesticide is manufactured)

RCRA
RCRA - Resource Conservation and Recovery Act
HSWA - Hazardous and Solid Waste Amendments
TCLP - Toxicity Characteristic Leaching Procedure
LDR or Land Ban - The Land Disposal Restrictions
TSDF - Treatment, Storage and Disposal Facility
LQG - Large Quantity Generator
SQG - Small Quantity Generator
BIF - Boiler and Industrial Furnace

FIGURE 2.1 EPA New England Inspector's Multimedia Checklist. *Courtesy USEPA*   *(Continued)*

TSCA/PCBs
TSCA - Toxic Substances Control Act
PCBs - Polychlorinated biphenyls
ML - Large PCB Mark

UST
UST - Underground Storage Tanks
OUST - Office of UST

WETLAND
CWA - Clean Water Act
404 - specific section of the CWA regulating the discharge of dredged or fill material into waters of the U.S., including wetlands

---

FIGURE 2.1 EPA New England Inspector's Multimedia Checklist. *Courtesy USEPA*

Other USEPA program-specific guidance for environmental audits include:

- Comprehensive Environmental Response, Compensation and Liability Act (CERCLA)[20]

- EPA Guidelines for Environmental Audits Under the Clean Water Act (CWA)[21]

- Emergency Planning and Community Right-to-Know Act (EPCRA)[22]

- Federal Insecticide, Fungicide and Rodenticide Act (FIFRA)[23]

---

[20] Protocol for Conducting Environmental Compliance Audits under the Comprehensive Environmental Response, Compensation, and Liability Act, EPA-305-B-98-009, December 1998 at http://infohouse.p2ric.org/ref/43/42045.pdf

[21] EPA Guidelines for Environmental Audits Under the Clean Water Act (CWA) at http://www.epa.gov/oecaerth/monitoring/programs/cwa/

[22] United States Environmental Protection Agency Enforcement and Compliance Assurance (2224-A) EPA305-B-01-002 March 2001 Protocol for Conducting Environmental Compliance Audits under the Emergency Planning and Community Right-to-Know Act and CERCLA Section 103 at http://www.epa.gov/compliance/resources/policies/incentives/auditing/apcol-epcra.pdf

[23] United States Enforcement and EPA 300-B-00-003 Environmental Protection Compliance Assurance September 2000 Agency (2221-A) Protocol for Conducting Environmental Compliance Audits under the Federal Insecticide, Fungicide, and Rodenticide Act (FIFRA) at http://www.epa.gov/compliance/resources/policies/incentives/auditing/apcol-fifra.pdf

- Resource Conservation and Recovery Act (RCRA) Hazardous Waste Generators[24]

- Safe Drinking Water Act (SDWA)[25]

- Toxic Substance Control Act (TSCA)[26]

In addition to the detailed guidelines and explanations offered by the protocols listed previously, it has been the author's personal experience that most regulatory staff appreciates the opportunity to help facilities comply with the regulations, making compliance inspections easier for the regulatory staff.

If the facility staff identifies the name of the facility to government regulatory staff, the regulatory staff answering the questions often takes "ownership" in the compliance process, and the inspectors will offer advice at inspections to make compliance simpler and easier. Remember these regulations are complex and confusing. It is much better to have the compliance inspector looking for words and phrases that support what the facility is doing to find compliance than to have the same words interpreted against the facility to prove noncompliance.

2.2.3.2 Legal considerations

The author is not a lawyer, and is not qualified to offer legal advice, but the USEPA publication, "Environmental Audit Program Design Guidelines"[27] contained the following excerpts which offer valuable insight into the federal government's use of voluntary environmental audit reports:

---

[24] Protocol for Conducting Environmental Compliance Audits of Facilities Regulated under Subtitle D of RCRA at http://www.epa.gov/compliance/resources/policies/incentives/auditing/apcol-rcrad.pdf
[25] Protocol for Conducting Environmental Compliance Audits of Public Water Systems under the Safe Drinking Water Act at http://www.epa.gov/compliance/resources/policies/incentives/auditing/apcol-sdwa.pdf
[26] Protocol for Conducting Environmental Compliance Audits of Facilities with PCBs, Asbestos, and Lead-based Paint Regulated under TSCA at http://www.epa.gov/compliance/resources/policies/incentives/auditing/tsca.pdf
[27] Environmental Audit Program Design Guidelines for Federal Agencies, USEPA, Office of Compliance and Enforcement Assurance (2261A) EPA 300-B-96-011, Spring 1997 at http://www.epa.gov/compliance/resources/policies/incentives/auditing/envaudproguidemas.pdf

Chapter 3: Legal Considerations

3.1 Overview

Designing and implementing an environmental audit program requires consideration of a number of legal issues. Chief among these is the protection of audit findings from premature disclosure. A comprehensive environmental audit typically accomplishes three objectives:

- Verify compliance/noncompliance with environmental regulations;

- Evaluate the effectiveness of environmental control systems; and

- Assess potential environmental liabilities from regulated and unregulated materials and practices.

To achieve these objectives, the audit findings must be candid, detailed, and accurate.

As such, environmental audits often describe actual or potential violations of law, unfavorable situations such as management deficiencies or inadequate staffing, or situations that do not constitute violations per se, but that nevertheless gives rise to potential environmental liabilities.

This kind of information can be used to the detriment of a facility or agency, and should be protected to the extent allowed by law. Public access to Federal agency documents and information in non-litigation situations is controlled by the Freedom of Information Act (FOIA)(5USC §552 et seq.). Once an agency audit report becomes final, it is an agency record and subject to disclosure through a FOIA request. As a result, the amount of time that an agency has to handle an environmental audit as an internal matter, free from outside scrutiny, is limited to that time between the conduct of the audit and the delivery of the audit final report.

Typically, a comprehensive environmental audit will contain information adverse to the audited facility. It is therefore important that the audit program be designed to provide for the protection of the audit findings from premature disclosure. Facility and agency personnel should have the opportunity to review and comment on audit findings, and develop a corrective action plan free from public scrutiny so that they can engage in free and frank discussions of regulatory opinion, interpretation, and applicability.

An understanding of privilege, as it pertains to audit reports, the FOIA law and process, and other legal considerations surrounding audit report handling and preparation will help in designing such a program. Please note that this discussion does not discuss document requests or subpoenas that arise from civil litigation.

Such requests must be handled through agency legal counsel on a case-by-case basis.

3.3 EPA Requests for Audit Reports

In the (its) 1995 audit policy, EPA reaffirms and clarifies its policy outlined in the 1986 audit policy to refrain from routine requests for audits. Eighteen months of public testimony and debate have produced no evidence that the Agency has deviated, or should deviate from this policy. In general, an audit which results in prompt correction clearly will reduce liability, not expand it. In addition, a review of the criminal did not reveal a single criminal prosecution for violations discovered as a result of an audit self-disclosed to the government. The 1995 policy states: "EPA will not request or use an environmental audit report to initiate a civil or criminal investigation of the entity. For example, EPA will not request an environmental audit report in routine inspections. If the Agency has independent reason to believe that a violation has occur red, however EPA may seek any information relevant to identifying violations or determining liability or extent of harm."[27]

Several states also have policies similar to the USEPA policy on voluntary audits. This should allay the concerns of facility owners that voluntary audit results may be used against them.

PLANNING AN ENVIRONMENTAL AUDIT

O nce a facility has decided to conduct an environmental audit, planning the audit is the next logical step. As discussed earlier, top-level management must be involved in these decisions, to ensure complete commitment at all levels. Further, in order for the audit to be successful, all management and staff must be committed to help conduct the audits and fully cooperate with the implementation of the recommendations.

Suggested Questions for Planning an Environmental Audit

In order to plan an audit, the facility management must first decide answers to the following questions:

- What are the reason(s) for the audit? As discussed in the Introduction, there are many reasons a company decides to conduct an environmental audit. Some are pro-active (e.g., the company owners or managers want to prepare for a transfer of property, maximize plant efficiency, explore cost-savings, minimize liabilities, prepare for application for a pollution control permit, or be prepared for future regulatory inspections). Other reasons may be reactive (a property transfer is imminent, a regulatory inspection is announced or has already been held, or there are complaints from the general public about nuisance conditions, such as noise, dust, etc.)

- What are the objective(s) to be achieved (programs or parts of programs to be audited). An environmental audit can be as simple as making sure some containers are properly labeled, or that a particular waste shipment is being taken away by a licensed hauler. However, another audit may be so complex as to encompass a huge facility that takes several days just to walk through. The facility must clearly identify all objectives before the audit starts.

- Which resources are required in terms of internal staff, outside consultants, lawyers, etc.? Ideally, company staff will be sufficiently trained to conduct the audit in-house. If not, it may be cost-effective to have them trained rather than hiring an outside consultant. This is true for many environmental audits. However, there are times when it is advisable to hire consulting firms or individuals with special training or expertise. As a creative alternative, the facility may offer a regulatory agency an opportunity to conduct a compliance inspection. While this sounds potentially dangerous, the regulatory representative who was asked to audit the facility will likely be using the inspection as a teaching or outreach opportunity.

- What is the overall scope (programs or portions of programs to be audited)? The author recommends making all potential audits as comprehensive as possible, because the facility may be visited by a regulatory compliance inspector from any program at any time. Federal funding for air resources and hazardous waste compliance inspections has been increasing steadily for the past several years, and the federal goals for these regulatory inspections have increased proportionately.

- To whom will the report will be available (in-house, regulatory agency, and/or public)? The audit will certainly be reviewed by the facility representatives, and in many cases, after considering any corrective actions, filed away for future reference. However, the audit may have been ordered under an enforcement action, so one or more regulatory agencies may be reviewing the audit report in detail. Also, some facilities make their audit reports available to the public, either through an enforcement order, or because the facility wants to reassure the public about their environmental compliance and the safety of the public. This practice is common when a facility is applying for an environmental permit, and they want to reassure the public and the regulatory agencies involved.

- Be ready to correct problems or potential violations that were discovered during the audit.

3.1 OBJECTIVES OF AN AUDIT

The objectives of an environmental audit are many and varied, but as discussed in the Introduction, examples include:

- Being prepared for environmental regulatory inspections or audits

- Responding to a legal requirement to conduct an audit as part of an enforcement action

- Keeping management and staff engaged in environmental compliance

- Ensuring compliance (keeping up with and staying ahead of the standards)

- Understanding costs of compliance (and noncompliance)

- Assuring customers that the facility is environmentally safe

- Projecting a responsible public image

- Minimizing liabilities

3.2 RESOURCES REQUIRED FOR AN AUDIT

The following resources must be available for an audit to be effective:

- Committed management structure (top-level Managers direct involvement and control).

- Trained staff and/or consultants, understanding all aspects of audited program(s), including:

  - Facility operation staff and managers

  - Knowledgeable engineers and technicians

  - Accountants, report editors

  - Attorneys

- Independent consultants in any or all of these roles, as necessary (see Chapter 4)

- Production and maintenance staff to interact cooperatively with auditors

- Monitoring equipment and trained technicians as needed

3.3 MANAGEMENT STRUCTURE

The potential permutations of facility management structure are as many and as varied as the types and sizes of each facility. The audit must have complete support and involvement from all owners, officers, CEOs, CFOs, and other management staff.

Oversight of the audit can take many forms, depending on the size and complexity of the facility, including but not limited to:

- Top management involvement through each step of the audit

- A smaller, delegated middle management group communicating major findings and decisions concerning the audit reporting to upper management

- Lower-level management groups supervising audit activities, communicating all findings and recommendations to middle- and upper-level management

All of these systems work, but no matter which management staff are chosen, any successful system requires direct involvement by upper-level management on all major decisions, keeping the top leaders aware of the findings, recommendations, and costs of implementation, so the top-level managers are responsive and, ultimately, responsible. This ensures complete ownership of the environmental ethic of the company by upper-level managers. Any management reporting structure that leaves out the direct involvement of top management is in danger of failure.

3.4 CHOICE OF STAFF AND/OR CONSULTANTS

In Chapter 4, there is a more detailed discussion regarding choosing the proper staff and/or consultants to conduct or participate in environmental audits, but it should always be recognized that many facility staff play important roles in environmental compliance at every facility. Planners of environmental audits need to be cognizant of staff's role, and involve staff as much as possible in most audits, for the purposes of improved morale, technical operational expertise, ownership of the audit process, and implementation of any corrective measures from the audit findings.

If consultants are needed, they should work closely with staff to maximize efficiency, and report to someone with a working knowledge of the facility.

3.5 MONITORING EQUIPMENT AND OPERATORS

Environmental compliance monitoring equipment can be very expensive to purchase or lease and operate, so those costs should be included in the budget for an environmental compliance audit, where applicable. For instance, the costs for purchasing, installing, and operating air monitoring equipment, like particulate matter (PM), carbon monoxide (CO), carbon dioxide (CO_2), and oxygen (O_2) probes for insertion into incinerator or boiler exhaust air stacks can be very expensive. Other expensive monitoring equipment includes radiation monitors for solid waste transfer stations, and laboratory equipment and operators for wastewater, hazardous waste, or air resources analysis.

At trial burns, for example, to verify compliance with hazardous waste and air resources permit conditions, or to apply for hazardous waste incinerator permits, the costs for leasing, installing, and monitoring the equipment for a few days can be hundreds of thousands of dollars per day, with trial burns sometimes lasting several days.

An example of some sophisticated air monitoring equipment is illustrated in Figure 3.1

3.6 LOGISTICS

Planning an environmental audit requires coordination of several resources at one time, including facility operation staff and managers, engineers and technicians, accountants, report editors, consultants, and attorneys (if necessary).

3.7 UNANNOUNCED REGULATORY COMPLIANCE INSPECTIONS

Unannounced regulatory compliance inspections are audits that cannot be planned by a facility, and a hypothetical course of events for a hazardous waste compliance inspection is provided separately in Chapter 10. However, it is prudent to have a contingency plan for logistics in responding to unannounced regulatory compliance inspections. An example is outlined below:

- Upon receiving word that an inspector is coming or has arrived:
  - Notify owner and managers (and staff) that an unannounced inspection is happening.

FIGURE 3.1 Air Monitoring Equipment. *Courtesy USEPA*

- Locate available conference room or other available space to host inspector(s).

- Invite inspector(s) and owner/management to meeting room for pre-inspection briefing (do not delay the inspector for more than a few minutes).

- Provide inspector(s) with all health and safety procedures needed while inspecting the facility.

- During the briefing, the inspector(s) will offer a proposed schedule, which most often starts with a review of records, followed by a facility walk-through, and ending with a final meeting, if desired by the facility (sometimes the inspector will ask to do the walk-through first, but this is uncommon).

- Have staff not attending the meeting check all areas for compliance. Depending on the inspector(s) proposed schedule, there may be several minutes or more for staff not in the briefing to prepare areas for inspection, because most inspectors will review records before inspecting the physical plant. If possible, during the briefing, have staff that are not in the briefing quickly check all areas for compliance and make any necessary corrections in areas that may be inspected, liking placing or correcting drum labels, putting covers or bungs on drums not in use, etc. (being careful not to cause violations during cleanup).

- After the initial briefing, most owner/manager staff will go back to their duties, leaving the inspector(s) to review records. Supply all requested records that are kept to the inspector(s), except for records you do not have, or records required to be kept in certain areas of the plant (like tank or container log books, daily and weekly monitoring reports, etc.).

- Accompany the inspector(s) on the walk-through, answering questions, and making appropriate staff available for questions you cannot answer. Make any records kept at the waste generation and management areas available upon request.

- After the walk-through, ask the inspector(s) for a post-inspection briefing, if not already requested. Invite owner/manager back to the meeting room. Discuss any questions, concerns, and potential violations with the inspector(s). If any of the inspector(s)' concerns or violation(s) can be corrected immediately, ask the inspector(s) to allow you to correct them immediately, and document them as corrected. If the violation can be corrected in the following few days, ask the inspector if you can email, fax, or deliver the corrections and have them be documented as corrected. (If a potential customer reviewing inspection reports sees that violations were corrected immediately, it obviously looks much better than having violations documented and taking up to 30 days to correct.)

- After the inspector(s) leave, hold an internal post-inspection briefing to discuss the events, the findings, what needs to be done to correct any problems, and finally, how the facility might better respond to future inspections.

Note that the time between the inspector(s)' announcement of the inspection and the walk-through is critical. If the staff knows all of the regulatory compliance rules and follows them, there should be little to check or correct when an unannounced inspection occurs. However, if this staff is trained properly and aware of the procedure during an unannounced inspection, there should be sufficient time to tidy up the areas as necessary and present areas that are in total compliance.

3.8 ANNOUNCED REGULATORY COMPLIANCE INSPECTIONS

The logistics for an announced regulatory compliance inspection are similar to the unannounced inspection, except that everyone has more time to prepare, and staff can do a better job of locating and organizing records, correcting potential violations, and tidying up the plant.

3.9 PLANNED FACILITY SELF-AUDIT

The logistics for an internal environmental audit have some of the same elements of a regulatory inspection, but a comprehensive environmental audit can encompass all environmental regulatory programs, and may also cover some voluntary programs as well.

In most facilities, an environmental audit will be conducted by several staff. For the sake of this text, the staff coordinating and conducting the audit will be called auditors. The example logistics plan provided is for a comprehensive, internal environmental audit, wherein the scope of the audit has been approved by the facility owner/management.

- Auditors' Pre-meeting. All auditors meet to:

  - Identify the organization of the audit team, including members and leadership structure (leader, recorder, staff liaison, etc.).

  - Discuss the scope of the audit approved by management, and agree which segments and duties will be handled by whom.

- Prioritize which programs should be audited in what order, based on potential exposure, liabilities, public attention, etc.

- Identify the applicable laws, rules, and regulations for each program and locate/prepare checklists for each program. (These checklists and customized questionnaires should be sent to operations/waste management staff in advance, if possible.)

- Identify voluntary program guidelines (if management approves) and locate/prepare checklists for these programs.

- Identify key staff for interviews and check their work schedules for availability.

- Review any previous audit reports, inspection reports, and other documents to build a history of compliance/noncompliance.

- Agree to an agenda for starting and conducting the audit.

- Agree on scope of post-audit meetings of auditors to discuss findings.

- Document findings from records reviews and site visits.

- Present audit findings and recommendations to owner/ management.

3.10 TIMING TASKS WITHIN AN AUDIT

It is important for the audit team to understand and plan for the sequence of activities at any audit. This is particularly true when certain segments of the audit are highly complex and may consume considerable more time than the other parts. For example, most chemical manufacturing facilities contain processes that often require a great deal more time for auditors to understand and complete an audit.

A useful and efficient way to plan audits with time-consuming segments is to use the Critical Path Method, wherein all necessary tasks are listed sequentially, and the concurrent (parallel) tasks that must be completed determine the ultimate duration of the audit. The critical path is the least amount of time required to complete the task (in this case, the audit).

A sample critical path schedule for an environmental audit is illustrated in Figure 3.2.

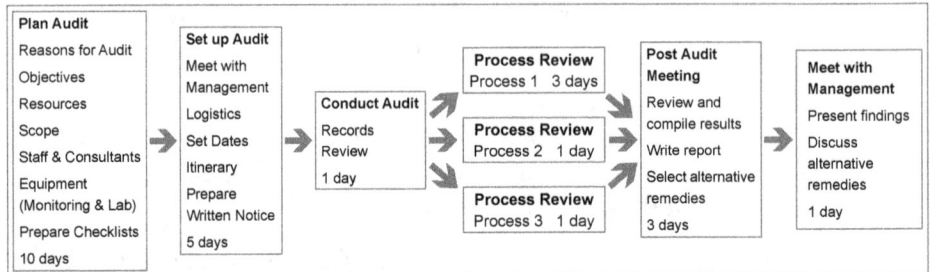

FIGURE 3.2 Critical Path Method for Environmental Audit. *Created by CVG Associates LLC May 2013*

Note that the three days necessary to complete Process 1 is the longest (3 days) and therefore part of the critical path in the audit process. Since the entire audit process had to wait for this audit process to occur, the projected critical path is 23 days.

If the audit had to be expedited, the rest of the tasks could be prepared awaiting the results of Process 1. This could reduce the time schedule to 21 days. An example of when this expedited approach might be used would be if there was an expectation of an unannounced regulatory audit in the immediate future.

3.11 LOGISTICS OF AN AUDIT

The logistics of an audit are very important, including timing, travel, meeting room space, and making sure relevant staff is available.

In the event of a simple one program check, logistics are fairly simple. However, in a comprehensive environmental audit of a large, complex facility, the travel plans can be complicated, especially if there are airline, train, or other mass transit reservations required. It is important that the travel timeframes be flexible, especially since the timing of some segments of the audit depend on completion of other segments, as illustrated in Figure 3.2.

The meeting times and durations also need to be flexible. It is recommended that the meeting rooms and interview rooms be blocked out during the entire anticipated audit timeframe, since a lack of meeting or interview space could result in delays in the final product.

Critical staff should also make sure they are available throughout the audit process, as their participation is vital to the audit's successful completion.

Next, Chapter 4 will address a critical issue: Who should conduct the audit?

WHO SHOULD CONDUCT THE AUDIT?

While choosing the appropriate auditors is sometimes a controversial decision, the decision of who should conduct an environmental audit is critical to the process. In general, the use of properly qualified and trained in-house staff to conduct these audits is almost always preferable to using outside consultants, because:

- In-house staff will generally be more familiar with the processes.
- In-house staff will generally have more ownership of the results.
- In-house staff will generally be more accountable than consultants.
- The costs are usually lower for using in-house staff.
- Staff will be properly trained for repeat or similar audits.

That said, there are circumstances that require the use of outside consultants instead of staff. For example:

- The audit may be the result of an enforcement action, and the party requiring the audit may stipulate that the audit be conducted by impartial, independent consultants.
- The company's internal staff may not be qualified to conduct part or all of the audit due to a lack of training, experience, or both.
- Some very technical portions of audits require highly specialized equipment, knowledge, and skills most often residing with a specialized

consultant. For instance, calibrations and trial burns for waste incinerators, boilers, and industrial furnaces require expensive, specialized equipment that needs to be operated by staff trained in their use. These calibrations and trial burns are usually only required every 5–10 years, so it is not economically practical for a company to retain this equipment or staff for infrequent activities.

- It may be more expensive to divert in-house staff away from their production work than to hire an outside consultant. Examples may include research and development, clean room production, and highly specialized staff with strict deadlines.

- Sometimes production staff is not completely cooperative with in-house EHS staff due to internal personality or staff (labor) management conflicts. An outside consultant can sometimes garner cooperation from staff that management might not.

4.1 PRODUCTION STAFF TO INTERACT COOPERATIVELY WITH AUDITORS

It is imperative for production staff at the facility to work closely and cooperatively with the auditors, as they have intimate knowledge of the machinery and methodology of making products and disposing of wastes. Involving production staff directly in the audit process also gives them ownership in the process and results.

4.2 HOW TO EFFECTIVELY INTERVIEW PRODUCTION STAFF

A properly trained auditor or inspector can put most employees at ease and obtain the information they need using a few simple methods:

- Give advance notice of employee interviews. Ideally, all employees to be interviewed should know the purpose of the interview, any documents needed, and the general questions in advance. These courtesies allow them time to formulate accurate and useful answers.

- Reassure them that you are visiting his/her site and interviewing them because they are the expert in that area and you only want to understand how they do their job. Tell them your goals are to understand the

FIGURE 4.1 Audit Interview. *Courtesy USEPA*

process and try to see if there are ways to make it easier, more efficient, or safer.

- Ask them questions in a comfortable place, preferably in or near their work area. Taking employees out of their normal work environment may make them nervous. In addition, they might want to use the equipment or waste containers to illustrate an answer. This can be noisy and result in distractions or interruptions, so be as patient as possible, and once they have shown you the area, suggest you retreat to a quieter area afterward to finalize your interview.

- Ask open-ended questions. Try to avoid questions that might elicit a "yes" or "no" response. Questions like, "What's this?" or "How does this work?" can get the person to open up and offer explanations. The phrase "Tell me about…" is very effective at putting people at ease, and getting them to offer information.

- Know when to stop an interview. If an employee is extremely nervous and you cannot get them to calm down, ask if they can talk about their answers to the questionnaire they received, and if that doesn't work, ask to discuss their answers later if you have any questions. If an employee is not helpful because they are hostile, sarcastic, or otherwise obstructive, stop the interview and speak with their supervisor to get the answers.

- Don't overemphasize violations. If you observe potential or real violations during the interview, you can point them out in a casual way, but don't dwell on the issue. Explain that the audit is nothing personal, just a means to improve the facility's compliance, and to avoid violations from a regulatory inspection.

- Don't talk down to the interviewee. Your knowledge of the regulations may exceed those of the person you are interviewing, but their knowledge of their job and what they believe the regulations require are the facts you are after. If you make the interviewer feel that you are somehow superior to them, they may not offer the information you need.

Case Study—Labor/Management Dispute Disrupts Environmental Audit

On a hazardous waste regulatory compliance inspection at a large facility, I arrived at the onset of a labor/management dispute. The union workers were not on strike at that time, but there was extreme tension between the company's management and the rank and file workers.

The Environmental Health & Safety representative (EHS) that was assigned to accompany me on the inspection warned me that the workers were very agitated, and a strike was imminent. After we had reviewed the documents required under the hazardous waste regulations, the EHS representative took me on a walk-through of the facility.

This was a fairly large facility, with over 30 points of hazardous waste generation, so there were a lot of partially filled 55 gallon drums in the satellite generation areas. As we walked around the facility, I observed workers intentionally spilling liquids from the containers onto the floor, removing or defacing labels from the containers, and removing bungs or covers from the containers. The first two of these acts (spilling and removing or defacing labels) represented clear violations of the rules, but the removal of covers and bungs was only a violation when the drum was not in use, and most of the drums were in use while I was there. I also observed the workers standing by and smiling while I informed the EHS manager about each violation they had caused.

After speaking with my central enforcement staff and our attorney, I was advised to document the violations, explain them to the EHS representative, and it was agreed we would postpone enforcement until the labor/management dispute was settled. The union workers went off strike a few days later, at which point I was invited back to see that the violations I had observed were corrected. I was able to close out the inspection with no serious violations.

In retrospect, if the company had indicated they had a labor/management dispute and it was unsafe to conduct the inspection, I would have asked my superiors if I could postpone the inspection until the dispute was settled.

Companies should inform inspectors whenever there is or may be an unsafe situation, and leave it up to the inspector whether they want to proceed. In general, most inspectors will not continue the inspection until the facility is safe and secure.

This illustrates the need to choose audit staff carefully, bearing in mind personal and professional relationships, education, training, professional certifications, personalities, and interpersonal/interview skills are all important factors in choosing staff.

After the decision is made concerning who should participate in the audit, an environmental audit is conducted (see Chapter 5).

CHAPTER 5

CONDUCTING THE AUDIT

This chapter is primarily devoted to conducting either program-specific or comprehensive environmental audits that are planned and voluntarily executed by a facility.

However, it is not unusual for facilities to experience unannounced environmental audits. Regulatory agencies sometimes conduct unannounced, program-specific, regulatory compliance inspections. These inspections cannot be anticipated or planned by the facility, so it is advisable for facilities to conduct voluntary facility audits in anticipation of regulatory inspections. Preparation for these unannounced regulatory compliance inspections is one of the prime drivers for voluntary environmental audits.

5.1 PRE-EMPTIVE ENVIRONMENTAL AUDITS

Facility owners/operators must weigh the cost of noncompliance from regulatory compliance inspections against the cost of pre-emptive voluntary environmental audits. In many cases, the cost of noncompliance can be much higher than the cost of pre-emptive audits.

Facility owners/operators have the option to conduct unannounced environmental inspections at their own facilities. These inspections would, after a period of time, follow the conclusion of company audits or regulatory compliance inspections, and be conducted to make sure the staff is properly executing the recommendations of the audit or inspection. However, these same unannounced environmental audits could be conducted in anticipation of a regulatory audit, especially if any regulatory audits are conducted at regular time intervals.

Audits should only be commenced after the proper planning for a program-specific or comprehensive environmental audit is complete, the correct resources have been selected, the proper regulatory research is conducted, and the logistics are set. This planning is outlined in Chapter 3, "Planning an Audit."

5.2 ANNOUNCED ENVIRONMENTAL AUDITS

Finally, a planned, pre-announced environmental audit is perhaps the most common audit, but the same degree of preparation is required regardless of whether the audit is announced or not.

The rest of this chapter is devoted to describing the events of a comprehensive environmental audit at a medium- to large-sized facility.

5.3 COMPREHENSIVE ENVIRONMENTAL AUDIT CHECKLIST

5.3.1 Pre-Audit Meeting

- The audit team (leader, recorder, staff liaison, writers, attorneys, etc.) meets to discuss the scope of the audit approved by management, and decide which segments and duties will be handled by whom. Leadership of each individual program should be assigned to an individual to ensure they have responsibility and authority to ensure complete coverage of each program.

- With management input, the team prioritizes, in order, which programs should be audited from first to last, based on potential exposure, liabilities, public attention, etc. These programs may be audited simultaneously if enough qualified staff is available, but, in most cases, the programs are audited sequentially.

- Regulatory checklists and customized questionnaires are sent to operations/waste management staff in advance of the audit. If management approves, the audit team identifies additional voluntary program guidelines (sustainability, waste management hierarchy, LEED, etc.).

- The audit team identifies the applicable laws, rules, and regulations for each mandatory and voluntary regulatory program and locates/prepares checklists for each program. These checklists should be written to include lessons learned from any previous audit reports, inspection reports, and other documents to account for the facility's history of compliance/noncompliance.

- In some industries, like chemical manufacturing, there are several complex chemical reactions used in making the product(s), and each of these processes need to be reviewed and understood by the auditors to ensure maximum efficiency, high product quality, minimum energy consumption, and minimum waste production (toxicity and volume). Analysis of these processes and their interrelationships is time-consuming and highly specialized. It may be advisable to retain a consultant to review the results of the audit, especially when the audit staff is not qualified to understand and interact with the production staff.

- Management and the audit staff discuss and agree to an agenda, audit protocol, and itinerary for the audit. An example of an agenda is included as Figure 5.1.

AGENDA FOR ENVIRONMENTAL AUDIT	
Days	**Activities**
1 & 2	Planning Meeting
3	Review of Project Documents
4	Contact with Facility Representatives
4	Prepare Audit Plan
5&6	Plan Questionnaires, Interview forms, and checklist
7&8	Develop Audit Protocol
9	Opening Meeting with Management, Auditors, and Staff
9	Document Review

FIGURE 5.1 Sample Audit Agenda. *Created by CVG Associates LLC*   *(Continued)*

9&10	Facility Walk-through and Interviews with Key Staff
11	Compilation of Facts
12	Closing Meeting
13	Draft Findings Report
14	Final Report
14	Identification of Findings and Corrective Action Alternatives
15+	Choose and Carry Out Corrective Actions
Weeks	Evaluation

FIGURE 5.1 Sample Audit Agenda. *Created by CVG Associates LLC*

A sample audit protocol is included as Figure 5.2.

A sample itinerary is shown in Figure 5.3.

- Key staff is identified for filling out checklists and participating in audit interviews, and their availability for the dates of the audit is verified.

- Meeting rooms are reserved and travel arrangements are made, if necessary.

SAMPLE AUDIT PROTOCOL
Applicability
Federal Legislation, Policies, Rules and Regulations
State and Local Laws, Rules and Regulations
Key Program(s) Compliance Requirements
Key Program(s) Terms and Definitions
Typical Records to Review
Typical Physical Features to Inspect
Interview Techniques
Checklists

FIGURE 5.2 Sample Audit Protocol. *Created by CVG Associates LLC*

SAMPLE ITINERARY FOR ENVIRONMENTAL AUDIT

Day 1

Time	Event
8:00 AM	Auditors arrive
8:15 AM	Pre-audit briefing with owner and or management
8:30 AM	EHS Manager briefs audit team on operation and answers questions
9:00 AM	General tour of facility to acquaint audit team with operations
12:00 PM	Lunch
12:30 PM	Records review of hazardous waste management program
3:00 PM	Site visit (includes interviews w/hazardous waste management staff)
4:30 PM	Return to meeting room to discuss findings
4:45 PM	Write draft findings
5:00 PM	Brief owner/manager on findings

Day 2

Time	Event
8:00 AM	Auditors arrive
8:15 AM	Records review of air resources program
8:30 AM	Site visit (includes interviews w/air resources staff and sampling)
1200 PM	Return to meeting room to discuss findings
1:00 PM	Lunch
1:30 PM	Continue site visit
4:00 PM	Return to meeting room to discuss findings
4:15 PM	Write draft findings
5:00 PM	Brief owner/manager on findings

Day 3

Time	Event
8:00 AM	Auditors arrive
8:15 AM	Records review of waste water treatment program
8:30 AM	Site visit (includes interviews w/wastewater staff and sampling)
1200 PM	Return to meeting room to discuss findings
1:00 PM	Lunch
1:30 PM	Continue site visit
4:00 PM	Return to meeting room to discuss findings
4:15 PM	Write draft findings
5:00 PM	Brief owner/manager on findings

FIGURE 5.3 Sample Audit Itinerary. *Created by CVG Associates LLC* *(Continued)*

Day 4	
Time	**Event**
8:00 AM	Auditors arrive
8:15 AM	Records review of petroleum and chemical bulk storage
8:30 AM	Site visit (includes interviews w/storage staff and sampling)
1200 PM	Return to meeting room to discuss findings
1:00 PM	Lunch
1:30 PM	Staff executes drill to activate contingency plan (SPCC)
3:00 PM	Return to meeting room to discuss findings
3:30 PM	Write draft findings
4:30 PM	Brief owner/manager on findings

Day 5	
Time	**Event**
8:00 AM	Auditors arrive
8:15 AM	Records review of all other environmental programs (including voluntary)
8:30 AM	Site visit (includes interviews w/storage staff and sampling)
1200 PM	Return to meeting room to discuss findings
1:00 PM	Lunch
1:30 PM	Write draft findings
2:30 PM	Compile all findings from entire audit
4:30 PM	Brief owner/manager on findings
5:30 PM	Hold final meeting with findings and recommendations

FIGURE 5.3 Sample Audit Itinerary. *Created by CVG Associates LLC*

5.3.2 Written Notice of Audit from Management

- Management sends out a written notice of audit to facility staff.

- Internal emails/calendar programs are used to block out time for audit for all key staff.

- Checklists and questionnaires are sent to key staff with deadlines for completion and estimated time windows when they will be interviewed.

5.3.3 Arrival at Site—Pre-Audit Briefing

- The audit team meets with the owner/manager, the EHS manager, and others (attorneys, report writers, etc.) to discuss the process of the audit and any remaining issues.

SAMPLE ANNOUNCEMENT OF ENVIRONMENTAL AUDIT

The CVG Company

Office Memo

To: All staff

From: Company President

Re: Environmental Audit

Date: June 2013

This is to let all staff know we will be conducting an internal environmental audit beginning the week of July1, 2013.

This comprehensive audit is being conducted to check our compliance with all of the environmental regulatory programs, including water, air, and solid & hazardous waste. The audit will also be reviewing our voluntary pollution prevention programs, including source reduction, reuse, recycling and electronic waste management.

This audit will be conducted by our in-house audit team, led by our EHS Manager. Please give them your full cooperation in this important effort.

Our EHS Manager and his staff will be requesting existing paperwork and written records necessary to verify compliance. Please give them your full cooperation as they seek and compile these records.

Some of you have received or will be receiving surveys from the audit team, including questions about how you do your jobs and your understanding of how to comply with the regulations. If you receive a questionnaire and have questions or concerns, please contact the EHS Supervisor at _____. The same people that receive these surveys may be interviewed by the audit team.

Thank you in advance for your cooperation. The goal of this audit is to make sure we are in compliance with all of the environmental standards, a goal I know you all share with me.

If anyone has questions about this audit, please contact the EHS Supervisor at _____.

FIGURE 5.4 Sample Company Notice for Audit. *Created by CVG Associates LLC*

- Members of the audit team that will be visiting the industrial areas of the site should be briefed on the site safety issues, and be trained to comply with all applicable safety regulations.

For an announced audit, all resources should be immediately available as arranged, including management and staff for pre-audit briefings, documents for review, location of environmental discharge locations, and waste management areas and facilities.

For an unannounced audit, the timing of all parts of the audit will generally take more time than an announced audit. The records review can be more challenging and time-consuming, because the records needed may be in several locations, including several locations on-site and even some storage off-site. Key staff may not be available due to other business demands, planned vacations, or sickness.

5.3.4 Records Review

The records review stage is a good time to ask questions regarding how the records are kept and where they are stored. There are almost always efficiencies to be gained in electronic records retention with the increased use of computers and improved data storage. Some regulations require records be stored on paper, but the trend is to accept more and more electronic copies. Federal and state regulatory agencies are beginning to encourage electronic record-keeping, data storage, and even report submittal, given the savings in paper and storage space, along with the ease and increased speed of retrieval.

For instance, most state governments encourage the submission of all annual hazardous waste generation reports electronically.

5.3.5 Walk Around

After the pre-audit meetings, it is usually the auditors' choice whether to review records or conduct the field visit. At an announced audit, much of the paperwork may have been reviewed already, so doing the field visit first may be more efficient or convenient. The auditor(s) may want to walk around the facility once for a general overview, followed by detailed visits to areas of concern. This can be more efficient if there are few areas where detailed inspections are necessary. If the facility has several areas that need detailed review, it might be more efficient to do include the walk-around with the detailed reviews conducted along the way.

5.3.6 Photographs/Videos

Photographs and/or videos should be taken wherever necessary to illustrate any potential issues or violations. They can also be used in audit debriefings to illustrate operations and issues to management and expert staff who were not present at the audit.

5.3.7 Property Boundaries

It is important to locate the facility's property boundaries, as some environmental compliance issues, like hazardous waste storage, wetlands delineation, forestry issues, and storm water may depend on the facility's boundaries and the distances of activities from these boundaries.

During the facility walk-around, the auditor(s) should verify compliance in every area, asking management, production, and waste management staff general and specific questions as necessary. If there is more than one process, and the processes are large, complex, and/or require specialized training or more time to audit, separate team members should break off to conduct those parts of the audit.

As mentioned earlier, the auditor(s) should interview the production and waste management staff with questions specific to the processes, but always verify the information independently. This is advisable because staff has developed their procedures from manufacturer's manuals, training, or experience, and their procedures may not be the most efficient method, or produce the highest level of compliance. These questions or interviews should always be friendly and open, encouraging staff and management input, including any ideas for process improvement or better waste management.

The detailed review of some processes can be highly involved, take a very long time, or both, particularly when the manufacturing processes are highly complex, involving chemical reactions with process loops, production of intermediates, and further chemical processes acting on those intermediates. Sophisticated processes should be reviewed by people who understand them thoroughly, because there may be efficiencies to be gained, like:

- Alternative feed chemicals that:
  - Are less expensive.
  - Are less toxic.
  - Produce less hazardous or nonhazardous by-products or wastes.

- Reactions that:
  - Require less energy.
  - Require less labor.
  - Require less steps.
  - Produce less waste (volume or toxicity).
  - Produce higher quality products.

For those process reviews that may take a longer time than other processes reviews, the audit team should schedule the review of these processes as early as possible in the audit, since these reviews may become part of the critical path as illustrated in Chapter 3 (Figure 3.1), and end up dictating the duration of the entire field work segment, and eventually the conclusion of the audit.

The audit team should agree on the scope of the post-audit meetings to:

- Discuss findings from all aspects of the audit to make sure they make sense.

- Document findings from:

  - Records reviews.

  - Site visits (walk-throughs).

  - Process reviews.

  - Interviews with staff.

- Present audit findings and recommendations to owner/management.

Chapter 6 explains how to write an audit report.

WRITING THE AUDIT REPORT

This chapter addresses the procedures for writing of a comprehensive environmental audit report. Depending on the facility audited, there will be findings from the audit of each applicable environmental program, and that could be as few as one, and as many as 21 to 23 environmental programs, listed below, depending on whether the company decides to evaluate voluntary programs, like the waste management hierarchy and the LEED program.

The first fifteen programs are regulated by the USEPA:

1. Generation of Hazardous Waste regulated under Resource Conservation and Recovery Act (RCRA) subtitle C.

2. Hazardous Waste Storage Tanks regulated under Resource Conservation and Recovery Act (RCRA) subtitle C (including subpart AA/BB/CC).

3. Treatment, Storage, and Disposal of Hazardous Waste regulated under Resource Conservation and Recovery Act (RCRA) subtitle C.

4. Management of Nonhazardous Solid Waste under Resource Conservation and Recovery Act (RCRA), subtitle D.

5. Emergency Planning and Community Right to Know Act (EPCRA).

6. Comprehensive Environmental Response, Compensation, and Liability Act (CERCLA).

7. Clean Air Act (CAA).

8. Clean Water Act (CWA).

9. Safe Drinking Water Act (SDWA).

10. Toxic Substances Control Act (TSCA).

11. Universal Waste.

12. Used Oil.

13. Pesticides—Federal Insecticide, Fungicide, and Rodenticide Act (FIFRA).

14. Management of other toxic substances (lead-based paint and asbestos).

The other less popular, but sometimes applicable programs are:

15. Low-level radioactive waste (LLRW).

16. Wetlands.

17. Stream Disturbance.

18. Floodplains and Floodways.

19. Lands and Forests.

20. Mining (mined land reclamation).

21. Fish and Wildlife.

Some voluntary programs are:

22. Waste Management Hierarchy initiatives; including:

    a. Source reduction

    b. Waste minimization

    c. Toxicity reduction

    d. Recycling

23. LEED (Leadership in Energy and Environmental Design).

Disclaimer: This is a comprehensive list of most environmental programs that are regulated by some governmental body, but the reader should be aware that there are other and sometimes varying legal requirements associated with facility compliance that can affect a facility's ability to operate.

Specific state and local rules may be different than the federal rules. For example, waste oil is regulated as a hazardous waste in some states, like Massachusetts and Connecticut, but not by many other states or the federal government. New York State regulates wastes with a concentration of greater than 50 parts per million (ppm) polychlorinated biphenyls (PCBs) as hazardous waste, but the federal government regulates wastes with the same concentration of PCBs under the Toxic Substances Control Act (TSCA).

6.1 COMPILING THE DATA

Before the report can be written, the audit team needs to compile all of the data from the records review and field visit, including:

- Observations of:
  - Compliance
  - Noncompliance
  - Areas of concern (need further research)
- Results of interviews with staff
- Field notes
- Findings from each individual program audit

6.2 ORGANIZING THE DATA

It is important to make sure all of the data is organized and easily accessible before writing the report. Each program audited will have its own set of data, and they should be segregated in a logical manner so as to be easily retrieved and inserted as necessary. Electronic storage of data can be very helpful by keeping the information in small, easily located, and transferrable packages.

Where there is a need to use hard copy (paper) reports, they should be filed in an orderly manner, so as to be easily located and retrieved.

6.3 WRITING THE REPORT

The next step after the compilation and organization of data is writing the audit report.

A properly prepared audit report should be:

- Thorough, covering all aspects of all applicable environmental programs.

- Accurate, with all information and interviews checked and verified.

- Well-organized, with a logical flow of information followed by conclusions.

- Detailed, with sufficient information given to illustrate complete analysis of all processes.

- Well-illustrated, with photographs, clearly understood graphics (charts, etc.).

- Well-written, using flowing language, with good grammar, sentence structure, and punctuation.

- Clear, understandable conclusions and specific alternatives and recommendations for correcting violations or areas of concern.

The findings from each individual program within the audit should be compared, and where findings from separate programs are related, they should be identified.

6.4 USE OF REPORT WRITERS

In Chapter 1, it was recommended that report writers be included as members of the audit team. While technical staff on the audit team is generally composed of experienced, technically qualified staff, the regulatory expert and production experts sometimes lack organizational or writing skills. If company management knows that the audit staff has good writing skills, it may be acceptable for these staff to prepare the final audit report. However, if management does not have confidence in the writing abilities of the audit staff, they should be prepared to engage experienced writers to help organize and write or edit the audit report.

6.5 CONSIDER THE AUDIENCE OF THE REPORT

Even if the facility management is comfortable with technical staff preparing an audit report, they should make sure any report being prepared for regulatory or public release is reviewed by one or more professional writer(s) or editors. This will help prevent any misunderstandings that might be caused by poor or unclear organization or writing. A poorly organized or incorrectly written report, containing bad grammar, misspelled words, or incorrect punctuation unnecessarily detracts from the message being conveyed by the facility and the technical team.

6.6 INDEPENDENT TECHNICAL REVIEW AND COMBINATION OF PROGRAM AUDIT REPORTS

Following the completion of the individual program audits, each section of the audit findings and conclusions should be reviewed by independent technical staff, if possible. If no peers are available to review, the audit team members should review each other's work for accuracy. If any of the individual program audits are highly complex or require special experience not available in-house, the company should retain a consultant or consultants with qualified experience. This is particularly applicable if the audit is required for property transfer, as a result of a consent order or consent decree,

After all program sections are reviewed for technical accuracy and completeness, preparation of the final comprehensive audit report should begin, including all programs.

6.7 GUIDANCE AND SAMPLE FORMAT

The format for the audit report should be simple, direct, and concise, and all of the findings and recommendations tabulated so they are easily read and understood.

The environmental regulatory audit guidance and checklists written and published by the USEPA listed at the end of Chapter 2 are reliable example resources, and should be utilized for format and content, if possible.

The general format for an audit report is illustrated below as Figure 6.1. A specific example of an audit report for hazardous waste is also included in the same exhibit.

EXHIBIT 6-1
EXAMPLE - COMPREHENSIVE ENVIRONMENTAL AUDIT REPORT FORM

Company Name _____

CompanyAddress _____

Facility (Division) Name (if different) _____

Facility (Division) Address (if different) _____

Contact Person _____

Title _____

Description of Facility _____

_____

_____

Products Made _____

_____

_____

Raw Materials Used _____

_____

_____

Standard Industrial Classification (SIC) Code(s) _____, _____, _____

To search for appropriate SIC codes, go to United States Department of Labor, Occupational Safety and Health Agency site: http://www.osha.gov/pls/imis/sicsearch.html

COMPLIANCE INSPECTION HISTORY

Compliance (Audit or Inspection) History for past 10 years) _____

Program _____

Results _____

Actions taken to correct _____

Use more pages if necessary.

Program _____

Results _____

Actions taken to correct _____

FIGURE 6.1 General format for Audit Report. *Created by CVG Associates LLC*   *(Continued)*

Use more pages if necessary.

Program

Results

Actions taken to correct

Use more pages if necessary.

VOLUNTARY WASTE MANAGEMENT IMPROVEMENT INTITIATIVES

Describe Pollution Prevention Initiatives (if any)

Source Reduction

Reuse of raw materials, byproducts or wastes

Recycling of materials or waste

Leadership in Energy and Environmental Design (LEED)

EMPLOYEE SUGGESTIONS FOR IMPROVED WASTE MANAGEMENT

Employee   Suggestion(s)

Employee   Suggestion(s)

Employee   Suggestion(s)

Employee   Suggestion(s)

Employee   Suggestion(s)

Use more pages if necessary

FIGURE 6.1 General format for Audit Report. *Created by CVG Associates LLC*   *(Continued)*

EXAMPLE
HAZARDOUS WASTE GENERATOR AUDIT FORM

EPA Hazardous Waste Generator ID # _ _ _-_ _ _ _ _ _ _ _

Activities that generate hazardous waste

Process _____

_____

Hazardous Waste(s) Generated (USEPA ID Codes and Description)

USEPA ID Code _____   Description _____

USEPA ID Code _____   Description _____

USEPA ID Code _____   Description _____

USEPA ID Code _____   Description _____

USEPA ID Code _____   Description _____

USEPA ID Code _____   Description _____

USEPA ID Code _____   Description _____

USEPA ID Code _____   Description _____

USEPA ID Code _____   Description _____

Use more pages as necessary

There are several categories of hazardous waste generators, listed below:

- Not a regulated Handler (exempt from hazardous waste regulations);
- Universal Waste Generator;
- Conditionally Exempt Small Quantity Generator;
- Small Quantity Generator;
- Large Quantity Generator

Note: In the Author's experience, very few facilities escape hazardous waste regulations completely, because many solid wastes generated in a business meet the definition of hazardous wastes. There is often significant confusion among environmental managers and industrial waste management staff concerning whether items like those listed below are hazardous wastes, because the very same wastes generated in households are excluded from regulation.

Hints: Offices generate hazardous wastes like:

- used fluorescent lamps;
- light ballasts;
- used computers;
- pesticides; and
- unusable oil-based and latex paints

For example, the USEPA guidance specifically identifying blown fluorescent lamps (bulbs) as hazardous waste was issued in 1996, and in a significant number of hazardous waste inspections conducted by me or my staff since 1996, a large number (50% – 80%) of the regulated community claimed they had no idea the rule existed.

FIGURE 6.1 General format for Audit Report. *Created by CVG Associates LLC*   *(Continued)*

Classification of Hazardous Waste Generators

Before a facility can be audited for compliance with hazardous waste regulations, it is necessary to determine if the facility generates hazardous waste, and if so, which generator category applies. The categories are listed above. An explanation of each category follows:

- **Non-regulated Hazardous Waste Handler.** The simplest and most lenient generator category is Non-regulated Hazardous Waste Handler. As discussed, above, this is difficult category to achieve because many commonly used products have hazardous characteristics when they are wasted.
- **Universal Waste Generator** Universal wastes are certain types of hazardous wastes that are generated on a regular basis by almost every business and industry (universally). Many of these wastes are so commonly generated that their associated hazards and dangers are often downplayed or ignored. Streamlined rules and regulations apply to the following, provided they are still in their original cases: batteries; pesticides; thermostats; and lamps. A Universal Waste Generator can still qualify as a Non-regulated Hazardous Waste Generator if the only hazardous wastes they generate are universal wastes and they manage those wastes as universal wastes.
- **Conditionally Exempt Small Quantity Generator** This category applies to companies who generate no more than 100 kilograms (220 pounds) of non-acute hazardous waste or 1 kilogram (2.2 pounds) of acute hazardous waste in a calendar month. They also cannot store more than 1,000 kilograms (2,200 pounds) of non-acute hazardous waste and no more than 1 kilogram (2.2 pounds) of acute hazardous waste at one time.
- **Small Quantity Generator** Companies who generate more than 100 kilograms (220 pounds) per month and no more than 1,000 kilograms (2,200 pounds) of non-acute hazardous waste or no more than 1 kilogram (2.2 pounds) of acute hazardous waste. They cannot accumulate more than 6,000 kilograms (13,200 pounds) of non-acute hazardous wastes and no more than 1 kilogram of acute hazardous waste.
- **Large Quantity Generator** Companies who generate more than 1,000 kilograms (2,200 pounds) of non-acute hazardous waste or more than 1 kilogram (2.2 pounds) of acute hazardous waste and store more than 1,000 Kilograms (2,200 pounds) of non-hazardous waste and more than one kilogram (2.2 pounds) of acute hazardous waste.

The rules and regulations for each of these categories are spelled out in the USEPA regulations. As an illustration, the rules and regulations for a Small Quantity Generator are outlined in checklist form below:

SMALL QUANTITY HAZARDOUS WASTE GENERATOR CHECKLIST

Requirement	Yes or No
Makes and documents hazardous waste determinations, including laboratory results	
Obtain a USEPA hazardous waste generator identification number (USEPA ID)	
Accumulate no more than 6,000 kilograms (13,200 pounds) of non-acute hazardous waste and 1 kilogram (2.2 pounds) of acute hazardous waste	
Arrange for shipment of all hazardous waste using a licensed hazardous waste transporter, accompanied by a hazardous waste manifest and LDR form	

FIGURE 6.1 General format for Audit Report. *Created by CVG Associates LLC* *(Continued)*

Must have hazardous waste shipped within 180 days if shipped < 200 miles	
May ship within 270 days if > 200 miles. If either of these time frames are exceeded, the generator is regulated as a large quantity generator	
Requirements for emergency management	
Post name and number of emergency coordinator next to telephone	
Post location of fire extinguishers and spill equipment next to telephone	
Post telephone number of fire department next to telephone	
Conduct training of staff on waste handling and emergency measures, and keep records of the content, when held, and who was trained	
Accumulation and storage requirements	
Containers in good condition and not leaking	
Wastes stored in containers of compatible materials	
Containers not in use are closed	
Containers are opened, handled, and stored to prevent leaks	
Each container must be marked with the words "Hazardous Waste" and with other words to identify its contents	
Wastes can be accumulated in accumulations areas 3 days or less after they are full They then need to be moved to a less than 90 day area or shipped off-site	
Less than 90 day storage area	
Date hazardous waste accumulation began clearly marked on each container	
Accumulation and storage requirements	
Containers in good condition and not leaking	
Wastes stored in containers of compatible materials	
Containers not in use are closed	
Containers are opened, handled, and stored to prevent leaks	
Each container must be marked with the words "Hazardous Waste" and with other words to identify its contents	
Each storage area must be inspected at least once per week, and itemize the 7 items listed immediately above	
If wastes remain in less than 90 day area more than 90 days, the facility is regulated as a treatment, storage facility	
Wastes stored more than 50 feet from property line	

FIGURE 6.1 General format for Audit Report. *Created by CVG Associates LLC*   *(Continued)*

Storage Tanks	
Meet all tank storage requirements in Parts 264 and 265	
Hazardous Waste Shipping	
Each shipment of hazardous waste must be hauled away by a licensed hazardous waste hauler	
Each shipment must be accompanied by a hazardous waste manifest	
Each manifest must be accompanied by a land disposal restriction form	

Note: Universal wastes generated at the facility do not have to be counted as part of the hazardous waste generated, provided the waste s managed in conformance with the universal waste rules.

Summary of audit findings

List the observed potential violations, accompanied by their regulatory citations below:

Concern or Violation	Citation	Recommended Remedy

Once the specific program checklists and lists of concerns and violations are completed, the results of all program audits chosen should be summarized in a final report, using the format similar to Pages 1, 2, and 3 in the form above.

FIGURE 6.1 General format for Audit Report. *Created by CVG Associates LLC*

NOTE *It is prudent to put as many of these forms in electronic format, to improve accuracy, efficiency and cost.*

CHAPTER 7

REVIEWING THE AUDIT FINDINGS AND MAKING CORRECTIONS AS NEEDED

After the audit is completed and the report submitted to management, the next logical step is to review the audit findings in detail and correct the report as needed.

In the author's experience, the benefits of conducting an independent review can be manifold, especially in the areas of source reduction, waste minimization, and recycling. The savings realized may even exceed the cost of the review in some cases.

7.1 FACTORS OF A SUCCESSFUL AUDIT

Audit reports can be highly complex and difficult to understand. The success of the audit depends on several factors, including:

- Accuracy of the audit (technical qualifications of the auditors).
- Thoroughness of the audit.
- Completeness of observations of physical operations at field visit.
- Diligence of the auditors in interviewing staff (completeness of interviews).
- Cooperation of production staff at field visit and interviews.
- Commitment of owner/management to provide necessary resources.

7.2 ACCURACY OF AUDIT (AUDITORS)

Some environmental programs use treatment and management technologies that are relatively simple and straightforward, with rules and regulations that are easily understood. However, many programs require highly specialized technologies that can be very complex. These same programs often are governed by regulatory language that is complex and often confusing, full of esoteric terms, and often result in conflicting interpretations. Auditing these programs can be very challenging, requiring individuals with specialized training and experience.

Companies need to be cognizant of these complexities, and make sure staff is fully trained and experienced in the programs being audited before proceeding. If unqualified or untrained staff are used for the audit, the quality of the audit may be compromised, and its purpose adversely affected. If the company lacks confidence in staff's abilities, they should hire an independent consultant(s) to assist in conducting the audit, and if possible, have the consultant(s) train the staff as the audit is conducted.

A few examples of very complex auditing issues are stack gas instrument calibration and measurement for air pollution control equipment on incinerators, boilers, and industrial furnaces; sampling and laboratory analysis of wastewater, hazardous waste, and mixed waste (hazardous waste mixed with low-level radioactive waste); and calibration and measurement of pesticide contamination.

7.3 THOROUGHNESS OF THE AUDIT

The facility needs to ensure that the audit is thorough, addressing all aspects of the program(s) being reviewed. The USEPA has prepared sample regulatory audit guidance that can and should be used as guidelines.[18-25]

---

[18] www.epa.gov/compliance/resources/policies/incentives/auditing/envaudproguidemas.pdf, op.cit.

[19] www.epa.gov/region2/capp/cip/r1mm99.pdf, op.cit.

[20] infohouse.p2ric.org/ref/43/42045.pdf, op.cit.

[21] www.epa.gov/oecaerth/monitoring/programs/cwa/, op.cit.

[22] http://www.epa.gov/compliance/resources/policies/incentives/auditing/apcolepcra.pdf, op.cit.

A good measure of the thoroughness of an audit is past regulatory inspections for the programs being audited, if any were held. Another excellent measure of thoroughness is a review of the pertinent regulations for the programs being audited. While some of these regulations are several pages in length, they contain all of the critical requirements for the program, and set the parameters for a regulatory inspector, should one come to visit.

7.4 COMPLETENESS OF OBSERVATIONS OF PHYSICAL OPERATIONS AT FIELD VISIT

Assuring completeness of observations at a field visit follows the same general guidelines as a thoroughness review. After reviewing any past inspection reports and the regulations, all of the physical plants associated with the program(s) being audited must be identified. The audit should address the inspection of each physical plant in sufficient detail to ensure compliance.

7.5 DILIGENCE OF THE AUDITORS IN INTERVIEWING STAFF

In Chapter 5, the text addresses suggested methods for interviewing production staff, giving ways to encourage an open, friendly dialogue. The information received from staff is critical to the audit, in that they work with the process more than anyone else at the facility, and if the physical part of the facility is designed correctly, their actions usually determine whether or not the regulations are being met and the facility is in compliance.

It is vital that the interviews are comprehensive, gathering all available information on how the operations are run, and the measures taken to ensure compliance.

---

[23] www.epa.gov/compliance/resources/policies/incentives/auditing/apcol-fifra.pdf, op.cit.

[24] www.epa.gov/compliance/resources/policies/incentives/auditing/apcol-rcrad.pdf, op.cit.

[25] www.epa.gov/compliance/resources/policies/incentives/auditing/apcol-sdwa.pdf, op.cit.

7.6 COOPERATION OF PRODUCTION STAFF AT FIELD VISIT AND INTERVIEWS

Staff may not always be fully cooperative with auditors, so it is important for the auditors to know when they are not receiving all the information or cooperation needed to complete the audit. The auditors may need to speak with the supervisors of employees who are not fully cooperative, either to gain the cooperation of the original employee, to speak with another employee who does the same duties, or to have the supervisor answer the questions. This is only a last resort, as most employees want to cooperate.

7.7 COMMITMENT OF OWNER/MANAGEMENT TO PROVIDE NECESSARY RESOURCES

There are occasions when additional resources are necessary to complete an environmental audit. Additional testing, interviews, and even independent experts may be needed to sort through issues that cannot be handled by the original audit team. Sometimes management will pull key members away from an audit team to accomplish another project. These issues can slow down or even stop an audit, and management needs to be made aware of the effect of their decisions. The timing of an audit may not be critical (as in a voluntary audit), so delays are not always time-sensitive. When this is the case, the audit can be finished later. However, if the audit is mandated and has a set deadline, management needs to be aware of the costs of any delay.

7.8 NEED FOR CLEARLY WRITTEN REPORT

As a supervisor and reviewer of hazardous waste, pesticides, and low-level radioactive waste regulatory inspections, the author supervised environmental engineers, chemical engineers, and geologists working in very highly specialized regulatory programs. Staff was charged with the responsibility to conduct comprehensive regulatory inspections (audits) and produce accurate and objective inspection reports. The author reviewed each of their inspection reports in detail for accuracy and completeness, and reviewed their recommended enforcement actions. All of the audits depended on completing and dealing with all the factors listed previously, and if any elements were missing or not done completely, the quality of the inspection report was diminished and had to be corrected.

While these inspection reports were objective from the inspectors' standpoints, they often needed corrections for grammar, spelling, sentence structure, etc. There were sometimes conflicts concerning the technical interpretation of regulatory language, especially given the complexity of the hazardous waste program and the confusing language that can be interpreted. In some cases, writing errors even changed the outcome of enforcement cases. Once the company received a notice of violation, the companies and their attorneys read the reports and listing of violations very carefully, and if errors were discovered, they often had a direct effect on the case; in some proceedings, either the prosecuting attorney would withdraw the case, or the judge would dismiss a case based on a single word.

7.9 VOLUNTARY AUDITS

As discussed in Chapter 6, if the audit is voluntary, and completed by staff, after the audit report is completed to the audit team's satisfaction, it is management's option to have the entire audit report independently reviewed by expert(s) who have extensive experience with environmental audits, comprehensive knowledge of the program(s) being audited, and excellent writing skills.

7.10 MANDATORY AUDITS

If the audit is required for legal or financial purposes, such as property transfer, required by consent order, consent decree, or judicial order, and company staff does not completely possess the necessary knowledge or skills to conduct a complete and accurate review, the company should strongly consider retaining an expert(s) as a consultant(s) to conduct a thorough technical review. If the owner/managers have confidence in the abilities of the staff that prepared the report, there may be no need for an independent expert review. This may require review by an independent attorney.

CHAPTER 8

SELECTING CORRECTIVE ACTIONS AS NEEDED

After an audit is completed, the owner/operator and company manager need to make decisions concerning what steps, if any, are necessary to correct shortfalls or violations found in the investigation.

A well-written audit report will either find no issues, or identify any problems that exist, and contain alternative solutions and recommendations to correct these problems, should they exist.

Company owner(s) and management have many decisions to make, such as:

- Evaluate the merits and possible pitfalls of the listed alternatives

- Consider any alternative solutions that may not have been listed in the report

- Decide what action(s) to take, evaluating the advantages and disadvantages of each

- Discuss the approach(es) with staff (or regulators if the audit was mandated)

- Direct staff and provide resources to correct the violations or other issues

The above list seems very concise, but the steps involved are far from simple. Each alternative can have many permutations and will not always yield the expected result(s). If the audit was mandated by a regulatory agency, the owner/operator should involve, or at least inform, the regulatory agency of its choices of steps to be taken, along with the reasons for taking those steps. This could prove invaluable because the regulatory agency will have some ownership in the process, and be more understanding and flexible should the alternative selected not yield a fully acceptable solution.

If the company is conducting the audit voluntarily, the recommendations are totally up to the owner/operator, and will hopefully be chosen for maximum compliance for the least investment. Often, a combination of voluntary initiatives, like source reduction, waste minimization, recycling, and energy optimization will make the audit cost-effective and worthwhile, in addition to the benefit of being ready for future compliance inspections, should they occur. The owner/operator should seek input from staff to increase ownership in the audit process and enhance the ultimate operation of the facility.

The owner/operator should document all corrective actions made from the audit findings and recommendations, along with the rationale for the decision.

In the author's experience, companies that placed environmental management high on their list of priorities in their business plans were highly prepared for environmental regulatory inspections, and talked at length about the savings they garnered while pursuing environmental excellence. An example of one such company follows.

Example of Company Management Making Environmental Compliance a High Priority

My staff and I took turns inspecting a research facility several times over a period of 12 years. The facility had a reasonable hazardous waste regulatory compliance record, with one exception. Some of the laboratories, especially those of some of the most successful inventors (chemists) were very cluttered and disorganized, where the vials and beakers holding wastes were not properly labeled, so it was impossible for the inspectors (and sometimes the chemist) to tell which vials contained reagents, feedstocks, by-products, intermediates, products, and wastes. This confusion on the part of the chemist resulted in documented violations, and eventually penalties.

The root cause of the problem was that the laboratory supervisors did not have the authority to order the chemists to keep their areas organized or to label their containers properly. In addition, the lab supervisor was not held responsible in any way, for these violation. After management was made aware of this shortfall, a system was set up wherein the laboratory supervisor would direct the chemists to label all their containers, and to document violations and levy "fines" in the form of reductions in pay of the chemist(s) who violated the rules. In addition, the chemists were informed in advance that any violations of the labeling rules that resulted in penalties levied as the result of inspections by the regulatory agencies would be the responsibility of the offending chemist(s). This met with much resistance from some of the chemists who had made important and valuable discoveries for the company. These chemists were informed by the company that the rules were firm, and as a result, compliance improved substantially.

A few of the "messier" chemists even attempted to mislead the laboratory supervisors concerning the contents of the containers. This was remedied by bringing in some other supervisors, along with experienced bench chemists, who sorted out the confusion and made certain all of the containers were labeled properly.

After an inspector found another series of labeling violations in one laboratory, at least one of the more productive chemists was fined and left the company because of the policy. The company accepted the loss, stating they felt the improvement in compliance and removal of the management challenges were worth more than potential future inventions.

8.1 SELECTING CORRECTIVE ACTIONS

This section is brief because selecting alternatives are usually business decisions, left to company owners and management.

In the case study above, the corrective action was to deal directly with uncooperative staff, setting an example for the rest of the company staff, and showing the commitment of the company to make environmental compliance a high priority. The company could have chosen to keep the staff and tolerate the noncompliance, but chose to clearly show their commitment to environmental compliance.

If the environmental audit was conducted as a result of an enforcement action, the regulatory authority will likely have input or approval authority over the chosen remedy(ies) for compliance. Sometimes the regulatory authority will allow the company to choose from one or more alternatives, provided the alternative ensures compliance.

If the company has complete latitude in their choices of alternatives, it should choose the alternatives that provide the best balance of economics and environmental compliance. Often, this will incorporate initiatives for source reduction, waste minimization, toxin reduction, recycling, and energy conservation.

All recommendations that have been resolved should include an explanation of how they were resolved.

CHAPTER 9

REVISING BUSINESS PLANS AND PROCEDURES AS NEEDED

As discussed in Chapter 8, after an environmental audit is completed, reviewed, and approved, companies must decide which actions, if any, need to be taken in reaction to the recommendations.

9.1 NO ACTION

In the rare event the environment audit doesn't require any changes to the physical plant or procedures, the company can continue with business as usual.

9.2 STAFF ACTION ONLY

If the audit recommends changes in staff actions (e.g., training, production methods, waste management procedures, inspection, etc.) or even changes in staff, the company needs to decide which course(s) of action will yield the best results from an economic and environmental compliance standpoint.

9.3 PHYSICAL PLANT IMPROVEMENTS OR ADJUSTMENTS

Environmental audits will sometimes recommend changes to the physical plant at a facility in order to meet environmental standards, such as larger capacity tanks, secondary containment, new or improved monitoring equipment, etc. These physical changes can vary in cost from a small investment to multi-million dollar upgrades.

In making each decision, the company needs to calculate if the investment is within its means, and if the company can stay economically viable. If not, the company needs to decide alternative courses of action, including modification of the business plan or shutting down that process or the whole company. Examples of other alternatives are to ship waste off-site or find a company that might be able to use the waste as a feedstock for one of their processes.

9.4 COMBINATIONS OF ALTERNATIVES

There are many combinations of the alternatives listed previously, and the company needs to decide which of these alternatives or combination of alternatives achieve the best compliance at the lowest cost.

9.5 AUTHOR'S PERSPECTIVE

As discussed earlier, companies make financial decisions constantly to maximize the financial viability of their business. Environmental compliance decisions can be critical, because they are often expensive, and can affect product quality and costs of production.

As a general rule, the author has observed that companies that face environmental regulations head on, and carefully analyze the intent of the lawmakers by looking at legislative history, have a better chance of succeeding in their business.

In over 30 years of developing hazardous waste regulations and enforcing those regulations by conducting regulatory hazardous waste inspections, the author observed hundreds of companies that were wasting money in vain attempts to comply with environmental regulations. Whenever the author could, the author would point out wasteful spending by well-meaning companies, with the best intentions of compliance, showing them the true intent of the regulations, and releasing them from many self-imposed, expensive solutions.

It is wise investment of time to study the reasons for laws before believing the rules and regulations are accurate reflections of those laws. Staff who write the federal regulations sometimes miss part or all of the intent of the lawmakers. The people who write the regulations are not necessarily licensed professional engineers, chemists, or people with experience in the business world they regulate. Some of these regulation writers have written rules, regulations, and guidance documents throughout their careers, and have never inspected or audited a facility. The author knows because he was one of those regulation writers. He wrote regulations for 6 years before he inspected one facility, and his perspective changed drastically after he conducted some inspections (audits) and understood what chemical manufacturing businesses actually do. As his understanding of their operations grew, he saw many flaws in the regulatory language, and relayed the problems back to the central regulatory staff for the state. Some of these comments resulted in changes to the regulations or to guidance used for future inspections.

It is highly unlikely that many regulation writers will visit facilities and gain the perspective the author did, so it is very important for the regulated community to realize that lack of real-world experience can lead to regulatory language that does not accurately reflect the laws, or more importantly, protect the environment.

If you have questions about rules and regulations and how to comply, a good, experienced inspector is often the best source for advice.

In Chapter 10, the reader will actually learn several tips about the best ways to deal with a regulatory inspection.

10

HOW TO RESPOND TO A REGULATORY ENVIRONMENTAL INSPECTION

The best way to prepare, or be prepared, for a regulatory inspection is to conduct a thorough and accurate internal environmental audit before the inspector(s) arrive.

This sounds inherently obvious, but the author has conducted hundreds of inspections where the company inspected was totally unprepared, and was cited for violations that could have easily been avoided at little or no cost. A pre-emptive environmental audit would have identified the violations and the corrections would have been made before the inspector(s) arrived, saving the companies violations and the attendant penalties.

10.1 AUTHOR'S EXPERIENCE WITH INSPECTIONS

After spending 20 years as a hazardous waste regulatory compliance inspector in New York State, the author conducted hundreds of hazardous waste and multi-media inspections, supervised several hazardous waste inspectors and pesticide inspectors, and personally knew every hazardous waste and pesticide inspector in the state. The author trained and certified a number of hazardous waste inspectors for federal inspection certification, and reviewed and approved thousands of inspection reports, referring the most serious violations to the central office for enforcement. Some severe violations, including potential criminal behavior, were referred to the State's Attorney General's Office for prosecution. In these cases the regulatory agency staff would work directly with the Attorney General's staff to prosecute the case.

The hazardous waste laws, rules, and regulations are arguably the most complex and confusing standards in the environmental field. While they were given several months of training, inspectors were sometimes confused about how to interpret some of these standards.

10.2 PERSPECTIVE ON INSPECTORS

The inspectors know the authority they have been given to inspect facilities, and want to accomplish these inspections as quickly and efficiently as possible, because they are assigned a large number of inspections to accomplish each year. They also know they are enforcing a highly complex program, and while they have an understanding of these standards, are aware that these standards are subject to more than one interpretation, depending on the particular circumstances at a facility.

Inspectors are human, and as such have egos that are sometimes bolstered by their authority to conduct inspections. Not unlike the police, they have the power to make it very difficult for companies by documenting violations and recommending enforcement actions. They also have the power to recommend leniency in enforcement of violations, citing facility cooperation, clean records, and other mitigating factors.

It is prudent for facility representatives to recognize the authority and powers possessed by these inspectors, and to understand that cooperation and politeness feeds the inspectors' egos. Conversely, disputing an inspector's opinions in a disrespectful manner can have a negative effect on their ego, possibly inviting more scrutiny and unnecessary strictness by the inspector(s).

As the author indicated earlier, most inspectors don't mind answering questions about the basis of a violation, or even calling their regulatory expert to find out, but if an inspector senses disrespect or encounters a negative or hostile attitude, they may become much more suspicious, strict, and unwavering in their interpretation of the rules.

This chapter is written to help facility owners, managers, and environmental managers respond to regulatory inspections. There are many variables and unlimited possible scenarios, but the advice offered here is universal to all inspections. The first scenario is that an inspection is announced and scheduled a day or more in advance, followed by the more probable unannounced inspection.

10.3 SCENARIO 1: THE UNANNOUNCED INSPECTION

In this scenario, inspector(s) are at the entrance to the facility or have just called and are on their way. This scenario is much more likely than an announced inspection (Scenario 2), because the USEPA prefers unannounced inspections and sets a goal for New York State of at least 90% unannounced hazardous waste inspections. The main differences in an unannounced inspection are on the part of the company. The inspector has the same exact work to do whether the inspection is announced or unannounced. The facility has no time to prepare; to conduct a pre-audit; to locate, prepare, or correct missing records; or to check labeling of containers or tanks.

10.4 SCENARIO 2: AN INSPECTION IS ANNOUNCED AND SCHEDULED

After a company receives notice that an environmental inspector has called and announced an inspection in the future, there are several steps that must be taken for a successful inspection. (For illustrative purposes, this example is for a hazardous waste inspection.)

Under either scenario, an inspection should follow the same general routine, based on the needs of the inspector.

10.5 SCHEDULE OF EVENTS IN A COMPLIANCE INSPECTION

10.5.1 After the Inspector Arrives

- Notify all persons involved about the inspection:
  - Owner/manager
  - Environmental, Health, and Safety (EHS) Manager

- Production staff
- Attorney (if necessary)

■ Call a meeting of all parties involved to discuss the inspection:

- Identify regulatory program(s) to be inspected and scope
- Locate and check all records that will be requested and reviewed
  - Waste determinations
  - Laboratory results
  - Permits (if any)
  - Daily, weekly, monthly, and annual records and reports
  - Shipping documents (manifests, land disposal restriction forms, and waste hauler reports)
  - Personnel training records
  - Contingency plans
  - Spill prevention, control, and countermeasure (SPCC) reports
  - Operating records
- Identify all facilities that may be inspected
  - Offices (used fluorescent light bulbs, ballasts, mercury containing equipment, expired paint, inks, photographic supplies, etc.)
  - Manufacturing/production facilities (waste generation points, satellite storage areas, and less than 90 day storage areas)
  - Waste storage areas (containers, tanks, etc.)
  - Waste neutralization/treatment facilities
  - Waste boilers and industrial furnaces
  - Waste incinerators
- Determine staff needed to be available to inspector(s)
  - Designated facility representative (EHS manager if one exists) to escort inspectors and answer questions

  – Staff trained to handle wastes

  – Production staff (as necessary)

  – Management/owner(s) and attorney(s) for post-inspection briefing

- Set up a comfortable meeting room for the inspector to review the records, conduct discussions, and hold a post-inspection briefing

If there is sufficient time before the inspection, the company should conduct a pre-inspection audit to identify and correct all potential violations before the inspector(s) arrive.

When the inspector(s) arrive at the appointed date and time, the facility representative should greet the inspector(s) and bring them to the conference room to meet the involved staff. If the facility manager/owner is available, they should meet the inspector(s) before the inspection starts (if possible).

The inspector(s) should not be kept waiting in the lobby or the meeting room for an unreasonable time (more than 5–10 minutes), as this may be perceived as unprofessional and might raise suspicion on the part of the inspector(s). One exception to this delay is a facility emergency, in which case the inspector(s) should be informed of the emergency and given an estimate of the delay. If the emergency is severe enough and the delay is significant, the inspector(s) may reschedule the inspection for another day.

The facility representative or the facility manager/owner should introduce the inspector and allow the inspector(s) to present their credentials and explain the reason for the inspection. This is a good time for the managers and/or staff to ask any general questions of the inspector(s), including whether the inspection is routine, and whether or not there are any pre-conceived concerns(for example, they may be responding to a tip or a complaint).

Once all questions and concerns are asked and answered at the pre-inspection interview, the inspector decides whether to first review some or all records, or conduct the site visit. Inspector's styles and methodologies differ, so they may express a choice to do either. The facility representative should be flexible on this point, as it is the inspector's prerogative.

10.5.2 Records Review Followed by Walk-Through

During the records review, it is advisable to produce all of the records requested, but it is not necessary or advisable to provide more documents than requested. Any volunteered additional materials might be filled out incorrectly or raise issues not previously considered by the inspector(s).

If the inspector(s) request records that the facility does not have and did not think was needed, ask the inspector(s) to explain why these records are necessary and the regulatory basis for the records. If the explanation is reasonable and is cited in the regulations, explain why the facility cannot produce the records, and whether there are alternative records. For example, the facility may not have prepared a contingency plan, but the spill prevention, control, and countermeasures (SPCC) plan may satisfy most or all of the requirements for the contingency plan. The facility may indeed be in violation of the regulations, so you would be wise to offer to create these records as soon as possible, starting immediately, and if possible, creating records retroactively if information is available in another report or format.

If the inspector(s) find that the records are incorrect, incomplete, or confusing, try to understand the issue and the basis for concern. After the inspector(s) explain this, ask them how these forms should be filled out, assure them you will have them filled out correctly in the future.

10.5.3 Site Walk-Through

During the facility walk-through, the facility representative(s) should stay with the inspector(s) at all times, answering any questions they might have to the best of staff's ability. Remember what was said about egos at the beginning of this chapter? Again, politely answer their questions in the most concise manner possible, without adding additional information. Anything added outside the parameters of their questions could raise concerns in other areas.

It may be unwise to show inspectors areas where waste is not generated or managed, because that may again raise concerns outside the scope of the inspection. If they ask to see a process or area where waste is not generated, exercise judgment concerning whether or not the inspector(s) need to see that area. They can be informed there are no wastes generated or managed there, but before doing so, consider their assignment and their egos. If they are curious or suspect wastes are generated or stored there, it may be wise to allow them to look than to have them suspect something.

10.5.4 Walk-Through First Followed by Records Review

If the inspector(s) request the walk-through before the records review, the same issues will be handled in a different order. Often an additional walk-through will be necessary when the initial walk-through is followed by the records review, but the results should be the same.

10.5.5 Final Meeting

After they have completed their site visit and seen all of the waste generation areas and waste storage areas (on big sites this may take several days), they will indicate when they are finished and, if requested, will meet and debrief the facility manager/owner.

The inspector(s) should be able to identify all of their concerns and potential violations (if any), and may ask additional questions based on observations. Answer their questions honestly and directly. "I don't know" is an acceptable answer, but it needs to be followed with "I will find out and get back to you ASAP (or by date and time certain)."

The inspector should be able to tell the facility representative all concerns and potential violations before they leave. Sometimes a concern may turn out to be a violation, and sometimes they turn out not to be an issue. This is a facility manager's/owner's opportunity to ask questions and get clarification on the potential violations and concerns, and where possible respond to some or all of them with feasible reasons for them. Remember the inspector(s) take their findings back to their office, prepare an inspection report, and submit it to their supervisor. The author was one of those supervisors, and can attest that some of the alleged violations were not true violations, due to a lack of evidence or a misinterpretation of the laws, rules, and regulations.

In New York State, after the inspector's supervisor signed off on the inspection report and potential violations, they are forwarded to a Central Office reviewer to further ensure accuracy and statewide consistency. During this review phase, even more potential violations can be removed based on inaccuracies or statewide policy. On the other hand, sometimes the violations are considered more serious on a statewide basis, and the enforcement recommendations are increased.

After all this review, the facility receives either a letter citing no violations; a warning letter usually giving a 30 day period to correct the violations; or a notice of violation, which means the facility could be subject to stricter enforcement, including a hearing by an Adjudicatory Law Judge, a possible consent order, and/or monetary fines. In very severe circumstances, when criminal activity is involved, there may be arrests, arraignments, court trials, and possible fines and/or incarceration.

10.6 NOTE FOR FACILITIES—BE A GOOD NEIGHBOR

Some inspections are initiated by citizen complaints. The facility may be noisy, dusty, emit noxious odors, or simply be in a location where it isn't wanted by certain neighbors.

In the author's experience, most facilities can gain favor with potential complaining neighbors using simple techniques. If your site releases dust on some occasions, it might be wise to offer free car washes to the affected neighbors. If the site is noisy, try to implement noise reduction technologies and consider modifying operating noisy equipment during daytime hours when people are awake.

Keeping good neighbors is a full-time job that pays dividends in the end.

10.7 WORD OF WARNING—DON'T REFUSE ACCESS TO INSPECTOR(S) WITHOUT GOOD CAUSE

There is never a convenient time to be inspected, but companies should think long and hard before they refuse entry to any environmental compliance inspector. Refusing entry to a regulatory inspector can be a very costly decision. For example, the maximum penalty for any hazardous waste violation is $37,500 per day, and refusing a hazardous waste inspector entry to any site is considered a major violation, supporting that kind of penalty.

The author experienced refusal of access on different occasions. In one case, after refusal of entry, a search warrant was obtained. A substantial penalty was assessed as part of several violations, including the refusal of entry. In two other cases, after the author explained the maximum penalty for refusal of entry, the facilities relented and offered access.

Note that if access to a property is refused, the inspector can normally obtain a search warrant within a few hours time, so refusal of entry only delays entry of the inspector for a very short time, and the penalty for refusal of entry is still enforceable.

If the facility is experiencing an emergency, like a fire, spill, or other event, notify the inspector(s) of the problem in sufficient detail of the dangers present, and the inspector(s) will likely either leave or wait until the emergency is under control. Be prepared to offer access as soon as the conditions allow entry, but you do not need to let them onsite during the emergency, unless they are duly authorized as official responders to a spill, fire, or other facility emergency.

CONCLUSION

Starting in the early 1970s, the United States Government started writing and passing the bulk of today's environmental laws, rules, and regulations. Many state and local authorities followed through, passing their own laws, rules, and regulations, and a bureaucratic maze of complex and confusing regulatory programs was created. The standards are voluminous, complex, and often confusing. Many of these standards can be read in multiple ways, and are subject to different interpretations by different people. There are a significant number of environmental areas and attendant regulatory programs, ranging from the most recognized programs of air resources, water resources, and solid and hazardous waste; to less recognized but often equally important programs such as chemical and petroleum bulk storage, high-level and low-level radioactive waste management, forestry, fish and wildlife, and many others.

This book represents the personal experiences gained by the author in the culmination of 20 years as an auditor of environmental compliance, working for New York State Department of Environmental Conservation, a state regulatory agency. Prior to those 20 years of working as a hazardous waste regulatory supervisor, trainer, and inspector in a field office for that state agency, the author helped develop the very same hazardous waste regulations, dealing with the USEPA and other states to come up with policies, procedures, and regulations that protected the environment by accurately reflecting the intent of the environmental laws passed by Congress and signed into law. That is not to say we all agreed on the content of these policies and regulations. There were many instances where in a state or several states would posit one position, and the USEPA would take another direction. In those cases, the USEPA would overrule the states' positions and develop the regulations as they felt best. Companies should be prepared to question unreasonable rules and regulations that do not seem to protect the environment, seeking to find the intent of legislators when then passed the laws.

As discussed and illustrated throughout this book, environmental audits are necessary and essential tools in helping companies succeed in the rapidly changing, dynamic business world. Companies that fail to make environmental compliance and its associated costs high priority elements of their business plans and budgets often place themselves in danger of financial failure.

This text prepares companies and individuals for environmental audits by regulatory agencies, and enumerates and illustrates over 20 environmental regulatory programs that might be encountered as part of any company's activities. Detailed explanations of the regulatory requirements are provided for some of the programs, and details of the other programs are provided in reference materials provided. Advice is given as to how to deal with the regulatory inspector, and how to prepare for their possible arrival. Beyond that, this book offers guidelines for environmental audits of homes and small offices, listing most environmental toxins encountered in people's daily lives, and offering advice on safer alternatives and ways to detoxify homes.

It is vital that each company familiarize themselves with all environmental programs applicable to their facility, or hire personnel that know these programs. A compliance check should be conducted on each program regularly, noting deficiencies and correcting them. These checks should sometimes be unannounced, placing the facility in a constant state of readiness for regulatory compliance inspections.

Environmental audits help ensure a company's well-being by ensuring compliance with all of those standards, and preparing companies and staff in the event of any unannounced regulatory audit by a government agency.

AVAILABLE GUIDANCE

United States Environmental Protection Agency, Guidance on Technical Audits and Related Assessments for Environmental Data Operations EPA QA/G-7 Final Office of Environmental Information Washington, DC 20460 EPA/600/R-99/080 at http://www.epa.gov/quality/qs-docs/g7-final.pdf

United States Environmental Protection Agency, Environmental Protection Enforcement Agency and Compliance Assurance EPA 300-B-96-011 (2261A) Spring 1997 at http://www.epa.gov/compliance/resources/policies/incentives/auditing/envaudproguidemas.pdf

United States Office of EPA 300-B-96-011 Environmental Protection Enforcement Agency and Compliance Assurance (2261A) Spring 1997 at http://www.epa.gov/compliance/resources/policies/incentives/auditing/envaudproguidemas.pdf

USEPA Environmental Auditing Policy Statement (51 CFR5004), updated December 1988 to EPA-305-B-009 at http://infohouse.p2ric.org/ref/43/42045.pdf

Incentives for Self-Policing: Discovery, Disclosure, 1-2 Correction and Prevention of Violations, December 22, 1995 (60 CFR6706) (1995 or final audit or self-policing policy) at http://www.epa.gov/compliance/resources/policies/incentives/auditing/finalpolstate.pdf

Website for U.S. Library of Congress at http://www.loc.gov/law/help/leghist.php

USEPA Emergency Management Guidance, Emergency Planning and Community Right-to-Know Act (EPCRA) Hazardous Chemical Storage Reporting Requirements at http://www.epa.gov/osweroe1/content/epcra/epcra_storage.htm

USEPA Publication EPA 745-B-00-002, February 2000, Emergency Planning and Community Right-to-Know Act(EPCRA) Section 313, Industry Guidance, petroleum storage and bulk storage facilities at http://www.epa.gov/tri/reporting_materials/guidance_docs/pdf/2000/2000petro4.pdf

USEPA Guidance on Agriculture, Federal Insecticide, Fungicide, and Rodenticide Act (FIFRA) at http://www.epa.gov/agriculture/lfra.html

Environmental Protection Agency's Incentives for Self-Policing: Discovery, Disclosure, Correction and Prevention of Violations; Notice Federal Register, Vol. 65, No. 70, Tuesday, April 11, 2000 at http://www.epa.gov/compliance/resources/policies/incentives/auditing/auditpolicy51100.pdf

Guidance prepared by the Federal Facilities Enforcement Office (FFEO), U.S. Environmental Protection Agency (EPA) for federal facilities to use as guidance in establishing and implementing environmental audit programs at http://www.epa.gov/compliance/resources/policies/incentives/auditing/envaudproguidemas.pdf

EPA New England Inpectors' Multimedia Checklist at http://www.epa.gov/region2/capp/cip/r1mm99.pdf

Protocol for Conducting Environmental Compliance Audits under the Comprehensive Environmental Response, Compensation, and Liability Act, EPA-305-B-98-009, December 1998 at http://infohouse.p2ric.org/ref/43/42045.pdf

EPA Guidelines for Environmental Audits Under the Clean Water Act (CWA) at http://www.epa.gov/oecaerth/monitoring/programs/cwa/

United States Environmental Protection Agency Enforcement and Compliance Assurance (2224-A) EPA305-B-01-002 March 2001 Protocol for Conducting Environmental Compliance Audits under the Emergency Planning and Community Right-to-Know Act and CERCLA Section 103 at http://www.epa.gov/compliance/resources/policies/incentives/auditing/apcol-epcra.pdf

United States Enforcement and EPA 300-B-00-003 Environmental Protection Compliance Assurance September 2000 Agency (2221-A) Protocol for Conducting Environmental Compliance Audits under the Federal Insecticide, Fungicide, and Rodenticide Act (FIFRA) at http://www.epa.gov/compliance/resources/policies/incentives/auditing/apcol-fifra.pdf

Protocol for Conducting Environmental Compliance Audits of Facilities Regulated under Subtitle Dof RCRA at http://www.epa.gov/compliance/resources/policies/incentives/auditing/apcol-rcrad.pdf

Protocol for Conducting Environmental Compliance Audits of Public Water Systems under the Safe Drinking Water Act at http://www.epa.gov/compliance/resources/policies/incentives/auditing/apcol-sdwa.pdf

Protocol for Conducting Environmental Compliance Audits of Facilities with PCBs, Asbestos, and Lead-based Paint Regulated under TSCA at http://www.epa.gov/compliance/resources/policies/incentives/auditing/tsca.pdf

Environmental Audit Program Design Guidelines for Federal Agencies, USEPA, Office of Compliance and Enforcement Assurance (2261A) EPA 300-B-96-011, Spring 1997 at http://www.epa.gov/compliance/resources/policies/incentives/auditing/envaudproguidemas.pdf

USEPA REGULATIONS

Solid and Hazardous Waste (Including Radiation)
http://www.epa.gov/osw/laws-regs/regs-non-haz.htm

For Hazardous Waste Generators
http://www.epa.gov/osw/hazard/downloads/tool.pdf

For Hazardous Waste Treatment, Storage, and Disposal Facilities
http://www.epa.gov/osw/hazard/tsd/permit/tsd-regs/tsdf-ref-doc.pdf

Water Resources
http://cfpub.epa.gov/npdes/npdesreg.cfm?program_id=0

Air Resources
http://epa.gov/air/oarregul.html

Medium Popularity Environmental Programs

Storm Water
http://cfpub.epa.gov/npdes/home.cfm?program_id=6

Environmental Cleanups
http://www.epa.gov/superfund/policy/remedy/sfremedy/regenfor.htm

Chemical/Petroleum Bulk Storage
http://www.epa.gov/oem/content/spcc/

Less Popular Environmental Programs

Pesticides
http://www.epa.gov/pesticides/regulating/laws.htm

Wetlands
http://water.epa.gov/grants_funding/wetlands/regulation.cfm

Stream Disturbance
http://water.epa.gov/polwaste/nps/mol1.cfm

Lands and Forests
http://www.epa.gov/agriculture/forestry.html

Mining (Mined Land Reclamation)
http://cfpub.epa.gov/npdes/indpermitting/mining.cfm

Wildlife
http://www.fws.gov/le/laws-regulations.html

HAZARDOUS WASTE COMPLIANCE CHECKLISTS

GENERAL INFORMATION AND CLASSIFICATION OF A FACILITY

Company Name _____

EPA ID# No. _____

Inspection Date _____

Part I

GENERAL INFORMATION AND CLASSIFICATION OF FACILITY

1. Identification of Hazardous Waste - 371   Yes   No

   A. Facility generates and/or stores hazardous waste
      on-site.   _____   _____

      (1) _____ Company has used knowledge of the waste to
                   determine if it is hazardous.

      (2) _____ The material has shown the characteristic of:

                   ( ) Ignitability (D001) - 371.3(b)

                   ( ) Corrosivity (D002) - 371.3(c)

                   ( ) Reactivity (D003) - 371.3(d)

                   ( ) Toxicity (D004 - 043) - 371.3(e)

(3) _____ The material is listed in the regulations as a hazardous waste from non-specific sources (F-Waste). 371.4(b).

(4) _____ The waste is listed in the regulations as a hazardous waste from specific sources (K-Waste). 371.4(c).

(5) _____ The material is listed in the regulations as an acute hazardous waste (P-Waste). 371.4(d)(5).

(6) _____ The material or product is listed in the regulations as a discarded commercial chemical product, off-specification species or manufacturing chemical intermediate (U-Waste). 371.4(d)(6).

(7) _____ The material is listed in the regulations as a waste containing PCBs (B-Waste). 371.4(e).

B. If the facility is a treatment, storage or disposal facility, have they:

_____ Submitted a Part A application.

_____ Should the Part A be modified by the Company? If so, explain.

_____

_____ Submitted a Part 373 permit application.

_____ Been granted a Part B permit.* expiration date: _____

_____ Been granted a Part 373 permit.* expiration date: _____

*Complete Appendix C - indicate compliance status with permit conditions.

C. _____ Has the facility signed a consent order to resolve violations found during a previous inspection?**

**Complete Appendix D and indicate compliance with *each* condition of the order.

2. Exemptions

A. Generator Exemptions

(1) _____ Not a regulated handler because:

(a) _____ Never generated any hazardous waste.

(b) _____ No hazardous waste generated within the last 3 years.

(c) _____ Company moved in _____ to _____.
                                    (date)   (location)

(d) _____ Company out-of-business.

(e) _____ Company sold to _____.
                                    (new owner)

(2) _____ Samples collected for testing - 372.1(e)(5).

(3) _____ Residues of hazardous waste in empty containers - 372.1(e)(6).

(4) _____ A hazardous waste which is generated in a product or raw material storage tank, transport vehicle or vessel, pipeline, or in a manufacturing process unit or an associated non-waste treatment manufacturing unit is not subject to regulation until it exits the unit in which it was generated, unless the unit is a surface impoundment, or unless the hazardous waste remains in the unit more than 90 days after the unit ceases to be operated, manufacturing, or for storage or transportation. - 372.1(e)(7)(i).

B. TSD Exemptions

(1) _____ Storage of hazardous waste that is generated on-site in containers or tanks for a period not exceeding 90 days. Other than the storage of liquid hazardous waste over the designated sole source aquifers - 373-1.1(d)(1)(iii).

(2) _____ Storage of liquid hazardous waste in containers (>185 gallons) or tanks generated on-site over the designated sole source aquifers for a period not exceeding 90 days. - 373-1.1(d) (1)(iv).

(3) _____ The on-site storage and treatment of hazardous waste by generators that generate less than 100 kilograms of hazardous waste in any calendar month and store less than 1,000 kilograms. - 373-1.1(d)(1)(v).

(4) _____ The storage and recycling of the recyclable materials identified in subparagraphs 371.1(g)(1)(iii) and (iv) of this Title - 373-1.1(d)(1)(vi).

(5) _____ The storage of the following recyclable materials is exempt from permitting provided that Subpart 374-1 is complied with. (NOTE: Subpart 374-1 requires that the facility also complies with selected sections of this Part.) - 373-1.1(d)(1)(vii):

(a) _____ recyclable materials used in a manner constituting disposal (see section 374-1.3);

(b) _____ hazardous wastes burned for energy recovery in boilers and industrial furnaces that are not regulated under section 373-2.15 or 373-3.15 of this Title (see section 374-1.8);

(c) _____ recyclable materials from which precious metals are reclaimed (see section 374-1.6);

(d) _____ spent lead-acid batteries that are being reclaimed (see section 374-1.7).

(6) _____ The recycling of hazardous wastes is exempt from permitting provided 373-2.2(c) (identification number), 372.4(b) (use of manifest system), 372.4(d)(1) (manifest discrepancies) and clause 373-1.1(d)(1)(viii)(d) are complied with. (Storage prior to recycling is not exempt under this subparagraph.) In addition: 373-1.1(d)(1)(viii):

(a) _____ This exemption is available to:

(1) _____ Commercial facilities that reclaim precious metals, as defined in 374-1.6 of this Title;

(2) _____ Mobile or transportable commercial facilities which operate on the generator's site, if a containment area, meeting the requirements of 373-2.9(f), is provided for the reclaiming facility and any associated, temporary container holding or storage area.

(b) ⎯⎯ This exemption is *not* available to any units, other than boilers and industrial furnaces, that burn hazardous wastes for energy recovery.

(c) ⎯⎯ Exempted processes that recycle the hazardous wastes listed in 2B(5)(a-d) must comply with Part 374 of this Title in lieu of the requirements specified in this subparagraph. (Note: Part 374 will require that the facility also complies with selected sections of this Part.)

(d) ⎯⎯ Owners or operators of facilities subject to RCRA permitting requirements with hazardous waste management units that recycle hazardous waste are subject to the requirements of sections 373-2.27, 373-2.28, 373-3.27 and 373-3.28 of this Part.

(7) ⎯⎯ The on-site treatment of hazardous waste, by the generator, in the same tanks or containers used for accumulation and storage is exempt provided the generator complies with Part 373- 1.1(d)(1)(iii) and (iv) and Part 372.2(c)(4). Any treatment or placement of hazardous waste in a manner that constitutes land disposal, as defined in subdivision 370.2(b), does not qualify for this exemption - 373-1.1(d)(1)(ix).

(8) ⎯⎯ Totally enclosed treatment facility - 373-1.1(d)(1)(xi).

(9) ⎯⎯ Elementary neutralization units or wastewater treatment units, as defined in Part 370 of this Title - 373-1.1(d)(1)(xii).

(10) ⎯⎯ Accumulation areas - 373-1.1(d)(1)(xiv).

(11) ⎯⎯ A transporter storing manifested shipments of hazardous waste in containers at a transfer facility for a period of ten calendar days or less - Complete Part VII - 373-1.1(d)(1)(xi).

3. Hazardous Waste Generation/Treatment/Storage/Disposal

    A. Describe only the activities that result in the generation of hazardous waste. Include manufacturing processes that generate hazardous waste. [Do not include hazardous waste treatment processes.]

_____

_____

_____

_____

_____

_____

_____

B. Describe any on-site hazardous waste treatment processes that result in the generation of hazardous waste (exempt and/or nonexempt). Include process diagrams if available.

_____

_____

_____

_____

_____

C. Identify the hazardous wastes that are on-site, the quantity of each, the storage method, the type and size of containers or tanks used and their location in the storage area. (Be as specific as possible.)

(1) Accumulation Areas [NOTE: Waste in accumulation areas must be included as part of the total quantity of waste stored on-site]:

_____

_____

_____

_____

(2) Container Storage Areas for CESQG, SQG, or Generator:

(3) Tank Storage Areas for CESQG, SQG, or Generator:

(4) Interim Status/Permitted Container Storage Areas:

(5) Interim Status/Permitted Tank Storage Areas:

(6) Treatment, storage or disposal units such as surface impoundments, landfills, waste piles, or incinerators:

4. Status Identification:

   A. Generator Status

      (1) _____ Conditionally Exempt Small Quantity Generator (CESQG) - generates less than 100 kg/mo of non-acute hazardous waste or 1 kg/mo of acute hazardous waste. Complete Part III - 372.1(f)(6), 371.1(f)(7).

      (2) _____ Small Quantity Generator (SQG) - generates more than 100 kg/mo but less than 1,000 kg/mo of non-acute hazardous, and accumulates no more than 6,000 kg of non-acute hazardous waste on-site. Complete Part IV - 372.2(a)(8)(iii).

      (3) _____ Generator - generates more than 1,000 kg/mo of non-acute hazardous waste or *generates more than 1 kg of acute hazardous waste in a calendar month*. Complete Part V - 372.2(a)(8)(ii).

   B. Treatment, Storage or Disposal Facility (TSDF)

      (1) _____ Hazardous waste is stored greater than 90 days.*,**

      (2) _____ Hazardous waste is received from off-site and not beneficially used, reused or legitimately recycled or stored.*

      (3) _____ Hazardous waste is treated on-site in non-exempt units.*

      (4) _____ Hazardous waste is disposed of on-site.*

   * (If checked Complete Part VI and/or appropriate Appendices)

   ** (Do not complete for generators only that have exceeded the 90 day storage limit.)

   C. Transporter Status

   Yes _____ No _____ Transporter operates a 10-day transfer facility.

   If Yes, Complete Part VII Permit No. _____

D. Universal Waste Handler

(1) _____ Small Quantity Handler - company accumulates no more than 5,000 kg total of universal waste at any time - Complete Appendix L.

(2) _____ Large Quantity Handler - Company accumulates 5,000 kg or more of universal waste at any time - Complete Appendix L.

(3) _____ Universal Waste Managed On-Site (list type and quantity).

_____

_____

_____

_____

E. RCRA Air Emission Rule (Subpart AA/BB/CC)

Is facility subject to RCRA Air Emission Rules (Subpart AA/BB/CC)?

_____ If Yes, Complete Appendix-X.

_____ If No, Please explain _____

_____

_____

_____

SMALL QUANTITY GENERATORS

Company Name ....

EPA ID# No. ....

Inspection Date ....

Part IV

SMALL QUANTITY GENERATOR (SQG)

Indicate: Indicate:

X Violations   X   Satisfactory

                                                NA Not Applicable

SQG - Small Quantity Generator - The generator who generates more than 100 kg/mo but less than 1,000 kg/mo of non-acute hazardous waste in a calendar month, and accumulates less than 6,000 kg on-site has complied with the following:

1. General Requirements - 372.2(a)

(a) .... The generator has made a determination as to whether or not his solid waste is a hazardous waste - 372.2(a)(2).

(b) .... The generator has obtained an EPA identification number - 372.2(a)(3)(i).

(c) .... The generator has not offered hazardous waste to transporters or to treatment, storage, or disposal facilities that have not received an EPA identification number - 372.2(a)(3)(ii).

(d) .... The quantity of non-acute hazardous waste accumulated on-site never exceeds 6,000 kg. - 372.2(a)(8)(iii)(<u>a</u>).

(e) .... Waste may be stored for up to 180 days unless the disposal facility is 200 miles or more away.

Storage up to 270 days is then allowed - 372.2(a)(8)(iv).

(f)   At all times there must be at least one employee on-site or on call with the responsibility for coordinating emergency measures - 372.2(a)(8)(iii)(e)(1).

(g)   The name and phone number of the emergency coordinator must be posted next to the telephone - 372.2(a)(8)(iii)(e)(2)(i).

(h)   The location of fire extinguishers and spill control material and, if present, fire alarm must be posted next to the telephone - 372.2(a)(8)(iii)(e)(2)(ii).

(i)   The telephone number of the fire department must be posted next to the phone unless the facility has a direct alarm - 372.2(a)(8)(iii)(e)(2)(iii).

(j)   The generator has ensured that all employees are thoroughly familiar with proper waste handling and emergency procedures - 372.2(a)(8)(iii)(e)(3).

(k)   The emergency coordinator or a designee have responded appropriately to any emergencies that have arisen - 372.2(a)(8)(iii)(e)(4).

2. Accumulation Area Requirements - 372.2(a)(8)(i)

(a)   The containers appear to be in good condition and are not in danger of leaking - 373-3.9(b).

(b)   Hazardous waste is stored in containers made of compatible materials - 373-3.9(c).

(c)   All containers except those in use are closed - 373-3.9(d)(1).

(d)   Containers holding hazardous waste must not be opened, handled or stored in a manner which may rupture the containers or cause them to leak - 373-3.9(d)(2).

(e) ⎯⎯ Containers are marked with the words "Hazardous Waste" and with other words that identify the contents of the containers - 372.2(a)(8)(i)(a)(2).

(f) ⎯⎯ Hazardous waste may be accumulated in excess of 55 gallons or 1 quart of acutely hazardous waste at or near the point of generation provided that Section 372.2(a)(8) (iii) requirements are met within 3 days, and the container holding the excess accumulation is marked with the date the excess amount began accumulating - 372.2(a)(8)(i)(b).

3. Container Storage Requirements - 372.2(a)(8)(iii)(<u>b</u>)

(a) ⎯⎯ The date upon which each period of accumulation begins is clearly marked and visible for inspection on each container - 372.2(a)(8)(iii)(<u>d</u>).

(b) ⎯⎯ Each container is marked with the words "Hazardous Waste" and with other words to identify the contents - 373-3.9(d)(3).

(c) ⎯⎯ The containers appear to be in good condition and are not in danger of leaking. (If containers are leaking, describe the type, condition, contents and number that are leaking or corroded. Be detailed and specific) - 373-3.9(b).

(d) ⎯⎯ Hazardous waste is stored in containers made of compatible materials - 373-3.9(c).
(*If not*, please explain.)

(e) _____ All containers except those in use are closed - 373-3.9(d)(1). _____

(f) _____ Containers holding hazardous waste must not be opened, handled or stored in a manner which may rupture the containers or cause them to leak - 373-3.9(d)(2). _____

(g) _____ The container storage area is inspected at least weekly - 373-3.9(e). _____

(h) _____ The generator complies with the following special requirements related to incompatible wastes - 373-3.9(g): _____

(1) _____ Incompatible wastes, or incompatible wastes and materials, are not placed in the _same container_, or in an unwashed container that previously held an incompatible waste or material unless the placement is conducted to prevent the following - 373-3.9(g)(1) & (2):

(a) _____ the generation of extreme heat or pressure, fire or explosion, or violent reaction - 373-3.2(h)(2)(i); _____

(b) _____ production of uncontrolled toxic mists, fumes, dusts or gases in sufficient quantities to pose a risk of fire or explosions - 373-3.2(h)(2)(ii); _____

(c) _____ production of uncontrolled flammable fumes or gases in sufficient quantities to pose a risk of fire or explosions - 373-3.2(h)(2)(iii);

(d) _____ damage to the structural integrity of the device or facility containing the waste - 373-3.2(h)(2)(iv); or _____

(e) _____ a threat to human health or the environment - 373-3.2(h)(2)(v). _____

(2) ⸺ Containers holding a hazardous waste that is incompatible with any waste or other materials stored nearby in other containers, piles, open tanks, or surface impoundments must be separated from the other materials or protected from them by means of a dike, berm, wall, or other device - 373-3.9(g)(3).

(i) ⸺ Special requirements for small quantity generators accumulating more than 185 gallons of liquid hazardous waste in storage areas *over sole source aquifers* - 373-1.1(d)(1)(iv)(g).

(1) ⸺ The container storage areas are within a secondary containment system designed and operated in accordance with the following - 373-1.1(d)(1)(iv)(f)(1):

(a) ⸺ The base under the containers must be free of cracks or gaps and sufficiently impervious to contain collected material until it is removed - 373-2.9(f)(1)(i).

(b) ⸺ The base must be sloped or the containment system otherwise designed and operated to drain and remove liquid unless the containers are elevated or protected from contact with accumulated liquids - 373-2.9(f)(1)(ii).

(c) ⸺ The containment system must have sufficient capacity to contain 10 percent of the volume of containers or the volume of the largest container, whichever is greater. Containers that do not contain free liquids are not considered in this determination - 373-2.9(f)(1)(iii).

(d) ⸺ Run-on is prevented unless the system has sufficient excess capacity over that required in (3) - 373-2.9(f)(1)(iv).

(e) ⸻ Accumulated waste and precipitation must be removed as necessary to prevent overflow - 373-2.9(f)(1)(v).

4. Tank Storage Requirements - 373-3.10(l)

(Complete Part IV-A)

5. Manifest, Reporting and Recordkeeping Requirement - 372.2(b) & (c)

(a) ⸻ Hazardous waste is shipped off-site with an accompanying manifest - 372.2(b)(5)(i).

If violation is checked, please provide details.

_____

_____

(b) List the frequency of shipments and the amount of waste per shipment.

_____

_____

(c) ⸻ The transporter has a valid Part 364 permit or is otherwise authorized to transport the waste to the designated facility - 372.2(b)(5)(ii).

(d) ⸻ The generator offers for shipment or ships hazardous waste to an authorized facility. - 372.2(b)(5)(iii).

If violation is checked, please provide details.

_____

_____

_____

_____

(e) ⸻ Each manifest is completed in accordance with the instructions found in Appendix 30 of Part 372 - 372.2(b)(1). [Indicate items in violation]

			Trans	Trans	
		Generator	1	2	TSDF

(1) ⎯⎯ Name of

(2) ⎯⎯ EPA ID No. of

(3) ⎯⎯ Mailing Address of

(4) ⎯⎯ Telephone No. of

(5) ⎯⎯ Manifest Document #

(6) ⎯⎯ The proper USDOT description.

(7) ⎯⎯ The appropriate: ⎯⎯ quantity, ⎯⎯ container number, ⎯⎯ container type, and ⎯⎯ waste type by units of weight or volume.

(8) ⎯⎯ Signed certification that the materials are properly classified, described, packaged, marked and labeled, and are in proper condition for transportation under regulations of the USDOT and NYSDEC.

(f) ⎯⎯ The generator has contacted the designated facility immediately after not receiving signed copies of manifests for wastes shipped off-site more than **35 days** ago - 372.2(c)(3):

⎯⎯ For these shipments, exception reports have been submitted after not receiving signed copies of manifests for **45 days** - 372.2(c)(3).

(g) ⎯⎯ The generator must distribute copies of the manifest as specified on the manifest form, postmarked within five (5) business days of the shipment date - 372.2(b)(3).

(h) ⎯⎯ For international shipments the generator has done the following - 372.2(b)(4)(i):

(1) ⎯⎯ The EPA and the Department have been notified 60 days prior to shipment of the hazardous waste destined for treatment, storage or disposal outside the United States - 372.5(c)(1).

(2) Primary exporters of hazardous waste must file with the Administrator and the Department no later than March 1 of each year, a report summarizing the types, quantities, frequency, and ultimate destination of all hazardous waste exported during the previous calendar year - 372.5(f)(1).

(i) The generator has complied with the requirements of Section 372.6 for interstate shipments - 372.2(b)(4)(ii).

(j) The generator has complied with the requirements for shipping by rail or water (bulk) found in Section 372.7 - 372.2(b)(4)(iii).

(k) The requirements of this section (Manifest Requirements) do not apply to hazardous waste produced by generators of less than 1,000 kg/m provided the following conditions are met - 372.2(b)(7):

(1) The waste is reclaimed under a contractual agreement pursuant to which - 372.2(b)(7)(i):

(a) the type of waste and frequency of shipments are specified in the agreement - 372.2(b)(7)(i)(a);

(b) the vehicle used for transporting the waste and delivering regenerated material back to the generator is owned and operated by the reclaimer - 372.2(b)(7)(i)(b); and

(c) the reclaimer complies with Part 364 - 372.2(b)(7)(i)(c).

(2) The generator records the following information for each shipment - 372.2(b)(7)(ii):

(a) the hazardous waste code and quantity of waste shipped - 372.2(b)(7)(ii)(a); and

(b) —— the date the waste is shipped - 372.2(b)(7)(ii)(b).

(3) —— The generator maintains a copy of the reclamation agreement for at least three years after termination or expiration of the agreement - 372.2(b)(7)(iii).

(l) —— A copy of each manifest has been kept for at least three years from the date the waste was accepted by the initial transporter - 372.2(c)(1)(i).

(m) —— A copy of each Exception Report must be kept for a period of at least three years from the due date of the report - 372.2(c)(1)(ii).

(n) —— A generator must keep records of any test results, waste analyses, or other determinations made in accordance with Part 372.2(a)(2) for at least three years - 372.2(c)(1)(iii).

(o) —— All records required under subdivision 372.2(c) were furnished upon request, or made available at a reasonable time for inspection - 372.2(c)(1)(iv).

(p) —— There is written communication that the designated treatment, storage or disposal facility is authorized for the hazardous wastes being offered for shipment, has capacity to accept such hazardous waste, and will assure the ultimate disposal method is followed - 372.2(b)(2)(i).

(q) —— There is written communication that the designated transporter is authorized to deliver the waste to the facility on the manifest - 372.2(b)(2)(ii).

6. Preparedness and Prevention - 373-3.3

(a) —— The facility is maintained and operated to minimize the possibility of a fire or explosion, or any unplanned sudden or non-sudden release of hazardous waste or hazardous waste constituents to air, soil, or surface water - 373-3.3(b).

(b) The facility must be equipped with the following, unless none of the hazards posed by waste handled at the facility could require a particular kind of equipment specified below - 373-3.3(c):

    (1) An internal communication or alarm system capable of providing immediate emergency instruction (voice or signal) to facility personnel - 373-3.3(c)(1);

    (2) A device, such as a telephone (immediately available at the scene of operations) or a hand-held, two-way radio capable of summoning emergency assistance from local police or fire departments or emergency response teams - 373-3.3(c)(2);

    (3) Portable fire extinguishers, fire control equipment, spill control equipment, and decontamination equipment - 373-3.3(c)(3); and

    (4) Water at adequate volume and pressure to supply water hose streams, or foam-producing equipment, or automatic sprinklers, or water spray systems - 373-3.3(c)(4).

(c) Facility communications or alarm systems, fire protection equipment, and spill control equipment are tested and maintained as necessary to assure their proper operation in time of emergency - 373-3.3(d).

(d) Personnel involved in hazardous waste operations have immediate access to an internal alarm or emergency communication device either directly or through visual or voice contact with another employee - 373-3.3(e).

(e) The owner or operator must maintain aisle space to allow the unobstructed movement of personnel, fire protection equipment, spill control equipment, and decontamination equipment to any area of facility operation in an emergency unless aisle space is not needed for any of these purposes - 373-3.3(f).

(f) The facility owner or operator has attempted to make the following arrangements as appropriate with local authorities for the type of waste handled at the facility and the potential need for the services of these organizations - 373-3.3(g)(1):

(1) Arrangements to familiarize police, fire departments and emergency response teams with the functions and layout of the facility - 373-3.3(g)(1)(i);

(2) Where more than one police and fire department might respond to an emergency, an agreement designating primary emergency authority to a specific police and a specific fire department, and agreements with any others to provide support to primary emergency authority - 373-3.3(g)(1)(ii);

(3) Agreements with State emergency response teams, emergency response contractors, and equipment suppliers - 373-3.3(g)(1)(iii); and

(4) Arrangements to familiarize local hospitals with the properties of hazardous waste handled at the facility and the types of injuries or illnesses which could result from fires, explosions or releases at the facility - 373-3.3(g)(1)(iv).

(g) Where state or local authorities decline to enter into such arrangements, the owner or operator has documented the refusal in the operating record - 373-3.3(g)(2).

UNIVERSAL WASTE

Company Name ....

EPA ID# No. ....

Region/Inspector ....

Inspection Date ....

<u>Indicate:</u>

X Violations

<u>Indicate:</u>

X   Satisfactory
NA Not Applicable

Appendix L

UNIVERSAL WASTE CHECKLIST

A. Standards for SQ Handlers and LQ Handlers of Universal Waste:

1. .... The universal waste is not disposed, diluted or treated on site by the handler - 374-3.2(b)/374-3.3(b).

2. .... A Large quantity handler of universal waste must notify the EPA Regional Administrator in writing about universal waste management unless he has already notified EPA of his hazardous waste management activities and has received an EPA Identification Number - 374-3.3(c).

3. Waste Management Requirements: 374-3.2(d)/374-3.3(d)

   (a) Universal Waste Batteries:

      (i) .... The handler must contain universal waste batteries in a container that is closed, structurally sound, compatible with the contents, and must lack evidence of leakage, spillage, or damage - 374-3.2(d)(1)(i)/374-3.3(d)(1)(i).

(ii) ⎯⎯⎯ A handler may conduct the activities such as, sorting, mixing, discharging, regenerating, disassembling, and removing of batteries from the product, as long as the battery cell is not breached and remains intact and closed - 374-3.2(d)(1)(ii)/374-3.3(d)(1)(ii).

(iii) ⎯⎯⎯ A handler who removes electrolyte from batteries, must determine whether electrolyte and/or other solid waste exhibit any characteristic of hazardous waste, and if so, it should be handled accordingly - 374.3.2(d)(1)(iii)/ 374-3.3(d)(1)(iii)

(b) Universal Waste Pesticides:

Does the waste pesticides meet the criteria of 374-3.1(c) to be universal waste? Yes ⎯⎯⎯ No ⎯⎯⎯

Please Explain: ⎯⎯⎯⎯⎯⎯⎯⎯⎯⎯⎯⎯

⎯⎯⎯⎯⎯⎯⎯⎯⎯⎯⎯⎯⎯⎯⎯⎯

(i) ⎯⎯⎯ The handler must contain universal waste pesticides in a container that remains closed, structurally sound, compatible with the contents and must lack evidence of leakage, spillage or damage - 374-3.2(d)(2)(i)/ 374-3.3(d)(2)(i).

(ii) ⎯⎯⎯ A tank that meets the requirements of section 373-3.10, except for subdivision 373-3.10(h)(3), 373-3.10(k) & (l), should be used to manage universal waste pesticides - 374.3.2(d)(2)(iii)/374-3.3(d)(2)(iii).

(c) Universal Waste Thermostats:

(i) ———— Any universal waste thermostat that shows evidence of leakage, spillage or damage must be contained in a container that is closed, structurally sound, and compatible with the contents - 374-3.2(d)(3)(i)/374-3.3(d)(3)(i).

(ii) ———— A handler may remove mercury-containing ampules from waste thermostats provided that the requirements (a)-(h) of this section are met - 374-3.2(d)(3)(ii)/374-3.3(d)(3)(ii).

(iii) ———— A handler who removes mercury-containing ampules from waste thermostats, must determine whether the mercury, or cleanup residues and/or other solid waste exhibit any characteristic of hazardous waste, and if so, it should be handled accordingly - 374-3.2(d)(3)(iii)/374-3.3(d)(3)(iii).

4. Labeling/Marking Requirement:

(a) ———— A container in which the batteries are stored *or* the waste batteries must be marked clearly with "Universal Waste - batteries" or "Waste batteries" or "Used batteries" - 374-3.2(c)(1)/374- 3.3(e)(1).

(b) ———— A container, tank or transport vehicle containing waste pesticides must be marked clearly with the label that was on the product *and* the words "Universal waste - Pesticides" or "Waste pesticides" - 374-3.2(e)(2) & (3)/374-3.3(e)(2) & (3).

(c) ⎯⎯ A container in which the thermostats are stored or a waste thermostat, must be marked clearly with "Universal Waste- Mercury thermostats" or "Waste Mercury thermostats" or "Used Mercury-thermostats" -374.3.2(e)(4)/374-3.3(e)(4).

5. Accumulation Time Limits:

   (a) ⎯⎯ The universal waste is not accumulated over a year from the date the waste is generated, or received - 374-3.2(f)(1)/ 374-3.3(f)(1).
   OR

   (b) ⎯⎯ The accumulation of universal waste for longer than one year is allowed, if the handler properly demonstrates that such accumulation is necessary to facilitate proper recovery, treatment, or disposal - 374-3.2(f)(2)/374-3.3(f)(2).

   (c) ⎯⎯ A handler must be able to demonstrate the length of time that the universal waste has been accumulated by marking the date, maintaining an inventory, or any other method - 374-3.2(f)(3)/ 374-3.3(f)(3).

6. ⎯⎯ A handler must inform all employees, who handle the universal waste, about the proper handling and emergency procedure - 374- 3.2(g)/374-3.3(g).

7. ⎯⎯ A handler of universal waste is prohibited from sending or taking universal waste to a place other than universal waste handler, or a destination facility - 374-3.2(i)(1)/374-3.3(i)(1).

8. ⎯⎯ When the universal waste is being transported off-site, by the handler or other transporter, the requirements of Part 364 must be met - 374-3.2(i)(2)/374-3.3(i)(2). (Note: Shipments less

than 500 pounds are exempt from Part 364, see 364.1(e)(3)(ii)).

9. A large quantity handler of universal waste must keep a record of each shipment of universal waste to and/or from the facility (i.e. a log, invoice, bill of lading, etc.). The records must be retained for at least three years from the date of the shipment - 374-3.3(j).

B. Standards for Universal Waste Transporter:

A universal waste transporter must comply with the requirements of 374-3.4(b) to (g).

C. Standards for Destination Facility of Universal Waste:

1. The owner or operator of a destination facility that recycles a particular universal waste without storing that waste before it is recycled must comply with 371.1(g)(3)(ii) requirements of this title - 374-3.5(a)(2).

2. The destination facility of universal waste is prohibited from sending or taking universal waste to a place other than universal handler or a destination facility - 374-3.5(b)(1).

3. The destination facility must keep a record of each shipment of universal waste to and/or from the facility (i.e. a log, invoice, bill of lading, manifest, etc.). The records must be retained for at least three years from the date of the shipment - 374.3.5(c).

LAND DISPOSAL RESTRICTIONS

Indicate:

X Violations

Indicate:

X   Satisfactory
NA Not Applicable

Company Name   ————————

EPA ID# No.   ————————

Region/Inspector   ————————

Inspection Date   ————————

Appendix A

LAND DISPOSAL RESTRICTIONS
(For SQG's, LQG's and TSD's that generate and/or store)

I.   Dilution Prohibited as a Substitute for Treatment

    A. ——— The generator or TSD, has not diluted in any way a restricted waste or the residual from treatment of a restricted waste unless: - 376.1(c)(1).

                Characteristic hazardous wastes are diluted (in a treatment system which treats wastes subsequently discharged to NYS waters) pursuant to a SPDES permit or for purposes of pretreatment under the Clean Water Act. [Dilution is permissible unless another method has been specified as the treatment standard in 376.4(c)(Five Letter Technology codes) or unless the waste is a D003 reactive cyanide wastewater or nonwastewater.)]

    B. ——— Combustion has been used to treat any of the hazardous waste codes listed in Appendix 54 (metal bearing wastes). Combustion is prohibited unless the waste, at the point of generation or after any bona fide treatment such as cyanide destruction prior to combustion can be demonstrated to comply with one or more of the following (unless otherwise specifically prohibited from combustion): - 376.1(c)(3)

- the waste contains hazardous organic constituents or cyanide at levels exceeding the constituent – specific treatment standard found in 376.4(j) of this Part;

- the waste consists of organic, debris - like materials (e.g., wood, paper, plastic, or cloth) contaminated with an inorganic metal - bearing hazardous waste;

- the waste, at point of generation, has reasonable heating value such as greater than or equal to 5000 BTU per pound;

- the waste is co-generated with wastes for which combustion is a required method treatment;

- the waste is subject to Federal and/or State requirements necessitating reduction of organics (including biological agents); or

- the waste contains greater than 1% Total Organic Carbon (TOC).

II. Testing, Tracking and Recordkeeping Requirements - 376.1(g)

   A. The generator has determined if the waste has to be treated before it can be land disposed - 376.1(g)(1)(i).

   The determination is based on: YES/NO

   a. Testing the waste

b. _____ Using knowledge of the waste

B. _____ For waste that does not meet the treatment standard: With the initial shipment of waste to each treatment or storage facility, the generator has sent a one-time notice to each receiving facility and placed a copy in the file. The notice must contain the following information: - 376.1(g)(1)(ii).

1. _____ EPA Hazardous waste number

2. _____ Manifest document number

3. _____ The waste is subject to the LDRs. The constituents of concern for F001-F005, and F039, and underlying hazardous constituents (for wastes that are not managed in a Clean Water Act (CWA) or CWA-equivalent facility), unless the waste will be treated and monitored for all constituents. If all constituents will be treated and monitored, there is no need to put them all on the LDR notice.

4. _____ The notice must include the applicable wastewater/ nonwastewater category and subdivisions made within a waste code based on waste-specific criteria (such as D003 reactive cyanide).

5. _____ Waste analysis data (when available).

6. _____ For hazardous debris, when treating with the alternative treatment technologies provided by subdivision 376.4(g): the contaminants subject to treatment, as described in paragraph 376.4(g)(2); and an indication that these contaminants are being treated to comply with subdivision 376.4(g).

NOTE: No further notification is necessary until such time that the waste or facility change, in which case a new notification must be sent and a copy placed in the generator's file.

C. For waste that meets the treatment standard at the original point of generation: With the initial shipment of waste to each TSD, the generator has sent a one-time notice to each TSD receiving the waste, and placed a copy in the file. The notice must include the following information: - 376.1(g)(1)(iii).

1. EPA hazardous waste number.

2. Manifest document number.

3. The waste is subject to the LDRs. The constituents of concern for F001-F005, and F039, and underlying hazardous constituents (for wastes that are not managed in a Clean Water Act (CWA) or CWA-equivalent facility), unless the waste will be treated and monitored for all constituents. If all constituents will be treated and monitored, there is no need to put them all on the LDR notice.

4. The notice must include the applicable wastewater/ nonwastewater category and subdivisions made within a waste code based on waste-specific criteria (such as D003 reactive cyanide).

5. Waste analysis data (when available).

6. The applicable certification.

NOTE: If the waste changes, the generator must send a new notice and certification to the receiving facility and place a copy in the file. Generators of hazardous debris excluded from the definition of hazardous waste under paragraph 371.1(d)(5) of this Title are not subject to these requirements.

D. ____ Wastes exempted from meeting treatment standards prior to land disposal: With the initial shipment, the generator must send a one-time notice to each land disposal facility receiving the waste. The notice must contain the following information: - 376.1(g)(1)(iv).

1. ____ EPA hazardous waste number.

2. ____ Manifest documents number.

3. ____ Statement: this waste is not prohibited from land disposal.

4. ____ Waste analysis data (when available).

5. ____ Date the waste is subject to the prohibition.

6. ____ For hazardous debris, when treating with the alternative treatment technologies provided by subdivision 376.4(g): the contaminants subject to treatment, as described in paragraph 376.4(g)(2); and an indication that these contaminants are being treated to comply with subdivision 376.4(g).

NOTE: If the waste changes, the generator must send a new notice to the receiving facility, and place a copy in their files.

E. Treatment of Prohibited Wastes in Containers or Tanks

____ For generators managing a prohibited waste in tanks, containers, or containment buildings, regulated under paragraph 372.2(a)(8) and treating that waste to meet applicable treatment standards, the following requirements have been met:

1. ____ Developed and followed written waste analysis plan which describes the procedures the generator will carry out to comply with the treatment standards - 376.1(g)(1)(v).

2. _____ The waste analysis plan has been based on a detailed chemical and physical analysis of a representative sample of the prohibited waste(s) being treated, and contains all information necessary to treat the waste(s), including the selected testing frequency - 376.1(g)(1)(v)(<u>a</u>).

3. _____ Kept the plan on-site in the generator's records - 376.1(g)(1)(v)(<u>b</u>).

4. _____ Wastes shipped off-site have complied with the notification requirements for restricted wastes meeting treatment standards - 376.1(g)(1)(iv)(<u>c</u>). [Complete Item II.C. pgs. A-4.]

F. Recordkeeping

1. _____ If a generator has determined whether a waste is restricted based solely on knowledge of the waste, all supporting data used to make this determination has been retained on-site in the generator's files - 376.1(g)(1)(vi).

2. _____ If a generator has determined whether a waste is restricted based on testing of the waste or an extract developed using the test method 1311, all waste analysis data has been retained on-site in the generator's files - 376.1(g)(1)(vi).

3. _____ If a generator has determined that he is managing a restricted waste that is excluded from the definition of hazardous or solid waste, or exempt from regulation, under 371, subsequent to the point of generation (including deactivated characteristic hazardous wastes managed in wastewater treatment systems subject to the Clean Water Act), the generator has placed in the facility's file a one-time notice stating such generation, subsequent exclusion or exemption from regulation and the disposition of the waste, in the facility's file. - 376.1(g)(1)(vii).

4.   Generators must retain on-site a copy of all notices, certifications, demonstrations, waste analysis data, and other documentation for at least three years from the date that the wastes were last sent to on-site or off-site treatment, storage, or disposal. This requirement applies to solid wastes even when the hazardous characteristic is removed prior to disposal, or when the waste is excluded from the definition of hazardous or solid waste, or exempted from regulation, subsequent to the point of generation - 376.1(g)(1)(viii).

G. Alternate Treatment Standards for Lab Packs.

If a generator is managing a lab pack containing hazardous waste and wishes to use the alternative treatment standards, The generator must submit a notice to the treatment facility with the initial shipment. The notice must contain the EPA hazardous waste codes, manifest document number, and the applicable certification. No further notification is necessary unless the wastes or receiving facility changes. For characteristic hazardous wastes (D001-D008 and D010-D043) underlying hazardous constituents need not be determined.
The recordkeeping requirements must be met - 376.1(g)(1)(ix).

H. Small Quantity Generators with Tolling Agreements.

For small quantity generators with tolling agreements, the following requirements - 376.1(g)(1)(x).

- For the *initial* shipment of such wastes, the generator has complied with the notification and certification requirements that apply for the wastes subject to the tolling agreement - 376.1(g)(1)(x). [Complete Items II.D, E, or F, pgs A-3 through A-5, as applicable, except for manifest requirements.]

- Small quantity generators must retain on-site a copy of the initial notification and certification, together with the tolling agreement, for at least three years after termination or expiration of the agreement - 376.1(g)(1)(x).

I. Hazardous Debris.

Generators or treaters who first claim that hazardous debris is excluded from the definition of hazardous waste under paragraph 371.1(d)(5) of this Title, (i.e., debris treated by an extraction or destruction technology provided by Table 1, subdivision 376.4(g), and debris that the commissioner has determined does not contain hazardous waste) are subject to the following notification and certification requirements: 376.1(g)(4).

1. A one-time notification must be submitted to the commissioner to include the following information: 376.1(g)(4).

a. The name and address of the authorized Part 360 facility receiving the treated debris - 376.1(g)(4)(i)(a).

b. A description of the hazardous debris as initially generated, including the applicable EPA or NYS Hazardous Waste Number(s) - 376.1(g)(4)(i)(b).

c. For debris excluded under subparagraph 371.1(d)(5)(i) of this

Title, the technology from Table 1, subdivision 376.4(g), used to treat the debris - 376.1(g)(i)(<u>c</u>).

2. ————— The notification must be updated if the debris is shipped to a different facility, and, for debris excluded under subparagraph 371.1(d)(5)(i) of this Title, if a different type of debris is treated or if a different technology is used to treat the debris - 376.1(g)(4)(ii).

III. Special Rules Regarding Wastes That Exhibit a Characteristic

A. ————— The initial generator of a solid waste have determined each EPA hazardous waste number (waste code) applicable to the waste in order to determine the applicable treatment standards under section 376.4 of this Part. For purposes of Part 376, the waste will carry the waste code for any applicable listing under section 371.4 of this Title. In addition, where the waste exhibits a characteristic, the waste will carry one or more of the characteristic waste codes under section 371.3, except when the treatment standard for the listed waste operates in lieu of the treatment standard for the characteristic waste, as specified in paragraph (2) of this subdivision. If the generator determines that their waste displays a hazardous characteristic (and is not D001 nonwastewaters treated by CMBST, RORGS, OR POLYM of subdivision 376.4(c), Table 1 of this Part), the generator must determine the underlying hazardous constituents (as defined in subdivision 376.1(b)(1) of this Part), in the characteristic waste - 376.1(h)(1).

B. ————— For a prohibited waste that is listed and also exhibits a characteristic, the treatment standard for the listed waste code will operate in lieu of the standard for the characteristic code, *provided* the treatment standard for the listed waste includes a treatment standard for the constituent that causes the waste to exhibit the characteristic. Otherwise the waste must meet the treatment standards for all applicable listed and characteristic codes - 376.1(h)(2).

C. _____ Prior to land disposal, all prohibited wastes which exhibit a characteristic have been treated to the treatment standards provided in 376.4 - 376.1(h)(3).

D. _____ Wastes that exhibit a characteristic are also subject to subdivision 376.1(g) requirements, except that once the waste is no longer hazardous, a one-time notification and certification must be placed in the generators' or treaters' files and sent to the commissioner. The notification and certification this is placed in the generators' or treaters' files must be updated if the process or operation generating the waste changes and/or if the Part 360 facility receiving the waste changes. However, the generator or treater need only notify the Department on an annual basis if such changes occur. Such notification and certification should be sent to the Department by the end of the calendar year, but no later than December 31 - 376.1(h)(4).

1. _____ The notification includes the following information - 376.1(h)(4)(i).

a. _____ The name and address of the Part 360 facility receiving the waste - 376.1(h)(4)(i)(a).

b. _____ A description of the waste as initially generated, including the applicable

EPA Hazardous Waste Number(s) and treatability group(s), and underlying hazardous constituents, unless the waste will be treated and monitored for all underlying hazardous constituents. In that case they do not have to be listed on the notice - 376.1(h)(4)(i)(<u>b</u>).

2. _____ The certification must be signed by an authorized representative and state the language found in subparagraph 376.1(g)(2)(iv)(<u>e</u>) - 376.1(h)(4)(ii).

3. _____ If the treatment removes the characteristic but does not treat underlying hazardous constituents, then the certification in 376.1(g)(2)(iv)(d) applies - 376.1(h)(4)(ii)(<u>a</u>).

IV. PCB Disposal

A. _____ All PCB wastes listed under Part 371 solely for their PCB content, are disposed of in accordance with the provisions of 40 CFR Part 761, except 376.4(f)(1).

1. _____ As listed in Part 371, waste B002, from any source other than a spill, may not be stabilized or mixed with any substance to conform with any provision of 40 CFR Part 761 regarding land disposal - 376.4(f)(1)(i).

V. Prohibition on Storage of Restricted Wastes - 376.5(a)

A. The storage of hazardous wastes restricted from land disposal is permitted provided that: - 376.5(a)(1).

1. _____ The generator has stored restricted waste in tanks or containers on-site solely for the purpose of the accumulation of such quantities of hazardous waste as necessary

to facilitate proper recovery, treatment, or disposal - 376.5(a)(1)(i).

2. The owner or operator of a hazardous waste treatment storage, or disposal facility has:

   a. _____ Only stored restricted wastes in tanks or containers for up to one year solely for the purpose of the accumulation of such quantities as necessary to facilitate proper recovery, treatment, or disposal - 376.5(a)(1)(ii).

   b. _____ Clearly marked each container or tank to identify its contents and the date each period of accumulation begins - 376.5(a)(1)(ii)(a).

   c. _____ Maintained in the operating record the contents and beginning accumulation date for each tank and container - 376.5(a)(1)(ii)(b).

   d. _____ Complied with all operating record requirements of 373-2.5(c) or 373-3.5(c) - 376.5(a)(1)(ii)(b).

B. _____ Unless the Department can prove that such storage was not solely for the purpose of accumulation of such quantities as necessary to facilitate proper recovery, treatment or disposal, the owner/operator of a treatment, storage or disposal facility may store restricted waste for up to one year - 376.5(a)(2).

C. _____ The owner/operator of a treatment, storage or disposal facility has stored restricted waste beyond one year and has proven that the storage was solely for the purpose of accumulation of such quantities of hazardous waste as necessary to facilitate proper recovery, treatment, or disposal - 376.5(a)(3).

D. _____ Liquid hazardous wastes containing PCBs at concentrations greater than or equal to 50 ppm have been stored at facilities that meet the requirements of 371 through 376 and 40 CFR 761.65(b), and have been removed from storage and treated or disposed of as required within one year of the date when such wastes were placed in storage - 376.5(a)(6).

TANK REQUIREMENTS

Part IV-A

TANK STORAGE REQUIREMENTS FOR SMALL QUANTITY GENERATORS

Indicate: Indicate:

X Violations   X   Satisfactory
                                                     NA Not Applicable

(a) General operating requirements

(1) _____ Hazardous wastes or treatment reagents must not be placed in a tank if they could cause the tank or its inner liner to fail - 373-3.10(1)(2)(ii).

(2) _____ Uncovered tanks must be operated to ensure at least 60 centimeters (2 feet) of freeboard, unless there is adequate containment - 373-3.10(1)(2)(iii).

(3) _____ Where hazardous waste is continuously fed into a tank, the tank must be equipped with a means to stop this inflow - 373-3.10(1)(2)(iv).

(4) _____ The owner or operator must mark all tanks with the words "Hazardous Waste" and with other words that identify the contents of the tanks - 372.2(a)(8)(iii)(d): 373-1.1(<u>d</u>)(1)(iii)(<u>c</u>)(<u>3</u>).

(b) Tank(s) are inspected each operating day for:

(1) _____ discharge control equipment (e.g. waste feed cutoff systems, bypass systems and drainage systems) - 373-3.10(1)(3)(i).

(2) _____ monitoring equipment (e.g. pressure and temperature gauges) - 373-3.10(1)(3)(ii).

(3) _____ level of waste in tank to ensure proper freeboard - 373-3.10(1)(3)(iii).

(c) Tank(s) are inspected weekly for:

    (1) ⎯⎯ corrosion or leaking of fixtures or seams - 373-3.10(1)(3)(iv).

    (2) ⎯⎯ erosion or obvious signs of leakage (e.g. wet spots or dead vegetation) of the construction materials of, and the area immediately surrounding discharge confinement structures (e.g. dikes) - 373-3.10(1)(3)(v).

(d) Tank closure

    (1) ⎯⎯ At closure, all hazardous waste must be removed from tanks, discharge control equipment and discharge confinement structures - 373-3.10(1)(4).

(e) Special tank requirements for ignitable or reactive waste

    (1) ⎯⎯ Ignitable or reactive waste is placed in a tank and the waste is stored, treated, rendered or mixed before or immediately after placement in the tank so that the resulting waste, mixture or dissolution of material is no longer ignitable or reactive - 373-3.10(1)(5)(i)(a)(1); and

    (2) ⎯⎯ The treatment, storage or disposal of ignitable or reactive waste in a tank is conducted so that it does not - 373-3.10(1)(5)(i)(a)(2):

        (a) ⎯⎯ generate extreme heat or pressure, fire or explosions violent reactions - 373-3.2(h)(2)(i);

        (b) ⎯⎯ produce uncontrolled toxic mists, fumes, dusts or gases in sufficient quantities to threaten human health - 373-3.2(h)(2)(ii);

        (c) ⎯⎯ produce uncontrolled flammable fumes or gases in sufficient quantities to pose a risk of fire or explosion - 373-3.2(h)(2)(iii);

(**d**)   damage the structural integrity of the device or facility containing the waste - 373-3.2(h)(2)(iv); or

(**e**)   through other like means threaten human health or the environment - 373-3.2(h)(2)(v); *or*

(**3**)   The waste is stored or treated in such a way that it is protected from any material or conditions that may cause the waste to ignite or react - 373-3.10(1)(5)(i)(**b**); *or*

(**4**)   The tank is used solely for emergencies - 373-3.10(1)(5)(i)(**c**).

(**5**)   The storage of ignitable or reactive waste in covered tanks complies with the National Fire Protection Association's (NFPA's) buffer zone requirements for tanks, contained in Tables 2-1 thru 2-6 of the "Flammable and Combustible Liquids Codes." - 373-3.10(1)(5)(ii).

(**f**) Special Tank Requirements for Incompatible Wastes

(**1**)   Incompatible wastes, or incompatible wastes and materials, are not placed in the same tank and hazardous waste is not placed in an unwashed tank which previously held an incompatible waste or material unless the mixture or commingling is conducted to prevent the following - 373-3.10(e)(6):

(**a**)   generation of extreme heat or pressure, fire or explosions, or violent reactions;

(**b**)   production of uncontrolled toxic mists, fumes, dusts, or gases in sufficient quantities to threaten human health;

(**c**)   production of uncontrolled flammable fumes or gases in sufficient quantities to pose a risk of fire or explosions;

(d) ⸺ damage to the structural integrity of the device or facility containing the waste; or

(e) ⸺ through other like means threaten human health or the environment.

(g) Secondary Containment Requirements for Tanks

Applicability: Small quantity generator must provide secondary containment system for tanks, at the time more than 185 gallons of liquid hazardous waste are accumulated, or at the time any liquid hazardous waste are accumulated in underground storage tanks - 373-1.1(d)(1)(iv) (g).

A. ⸺ Secondary containment systems must be designed, installed and operated to prevent any migration of wastes or accumulated liquids out of the system to the soil, groundwater, or surface water at any time during the use of tank system - 373-3.10(d)(2)(i).

B. ⸺ Secondary containment systems must be capable of detecting and collecting releases of accumulated liquids until the collected material is removed - 373-3.10(d)(2)(ii).

C. At a minimum, the containment system is:

1. ⸺ constructed of or lined with materials that are compatible with the wastes to be placed in the tank system and must have sufficient strength and thickness to prevent failure due to pressure gradients (including static head and external hydrological forces), physical contact with the waste to which they are exposed, climatic conditions, the stress of installation, (including stresses from nearby vehicular traffic) - 373-3.10(d) (3)(i);

2. _____ placed on a foundation or base capable of providing support to the secondary containment system, providing resistance to pressure gradients above and below the system, and preventing failure due to settlement, compression, or uplift - 373-3.10(d)(3)(ii);

3. _____ provided with a leak detection system that is designed and operated so that it will detect the failure of either the primary and secondary containment structure or any release of hazardous waste or accumulated liquid in the secondary containment system with 24 hours, or at the earliest practicable time if the existing detection technology or site conditions will not allow detection of a release within 24 hours - 373-3.10(d)(3)(iii); and

4. _____ sloped or otherwise designed or operated to drain and remove liquids resulting from leaks, spills, or precipitation. Spilled or leaked waste and accumulated precipitation must be removed from the secondary containment system within 24 hours, or in as timely a manner as is possible to prevent harm to human health or the environment, if removal of the released waste or accumulated precipitation cannot be accomplished within 24 hours - 373-3.10(d)(3)(iv).

D. Secondary containment for tanks includes one or more of the following devices: 373-3.10(d)(4).

1. _____ a liner (external to the tank) [Complete Item E1];

2. _____ a vault [Complete Item E2];

3. _____ a double-walled tank [Complete Item E3]; or

4. _____ an equivalent device as approved by the Commissioner.

E. In addition to Items A through D above, secondary containment systems must meet the following requirements:

1. External liner systems must be - 373-3.10(d)(5)(i):

(a) _____ designed or operated to contain 100 percent of the capacity of the largest tank or the volume of all interconnected tanks, whichever is greater, within its boundary - 373-3.10(d)(5)(i)(a̲);

(b) _____ designed or operated to prevent run-on or infiltration of precipitation into the secondary containment system unless the collection system has sufficient excess capacity to contain run-on or infiltration. Such additional capacity must be sufficient to contain precipitation from a 25-year, 24-hour rainfall event - 373-3.10(d)(5)(i)(b̲);

(c) _____ free of cracks or gaps - 373-3.10(d)(5)(i)(c̲).

(d) _____ designed and installed to completely surround the tank and to cover all surrounding earth likely to come into contact with the waste if released from the tanks (i.e. capable of preventing lateral as well as vertical migration of the waste. For onground tanks, the external liner system must also encompass the bottom of the tank) - 373-3.10(d)(5)(i)(d̲);

(e)   external concrete liners must be constructed with chemical-resistant water stops in place at all joints (if any) - 373-3.10(d)(5)(i)(e); and

(f)   external concrete liners must be provided with an impermeable interior coating that is compatible with the stored waste and that will prevent migration of waste into the concrete - 373-3.10(d)(5)(i)(f).

2. Vault systems must be - 373-3.10(d)(5)(ii):

(a)   designed or operated to contain 100 percent of the capacity of the largest tank or the volume of all interconnected tanks, whichever is greater, within its boundary - 373-3.10(d)(5)(ii)(a);

(b)   designed or operated to prevent run-on or infiltration or precipitation into the secondary containment system unless the collection system has sufficient capacity to contain run-on or infiltration. Such additional capacity must be sufficient to contain precipitation from a 25-year, 24-hour rainfall event - 373-3.10(d)(5)(ii)(b);

(c)   constructed with chemical-resistant water stops in place at all joints (if any) - 373-3.10(d)(5)(ii)(c);

(d)   provided with an impermeable interior coating or lining that is compatible with the stored waste and that will prevent migration of waste into the concrete - 373-3.10(d)(5)(ii)(d).

(e) ⸻ provided with an exterior moisture barrier or be otherwise designed or operated to prevent migration of moisture into the vault, if the vault is subject to hydraulic pressure - 373-3.10(d)(5)(ii)(<u>f</u>); and

(f) ⸻ provided with a means to protect against the formation of and ignition of vapors within the vault, if the waste being stored or treated - 373-3.10(d)(5)(ii)(<u>e</u>):

(1) meets the definition of ignitable waste under section 371.3(b); or

(2) meets the definition of reactive waste under section 371.3(d) and may form an ignitable or explosive vapor.

3. Double-walled tanks must be - 373-3.10(d)(5)(iii):

(a) ⸻ designed as an integral structure (i.e., an inner tank within an outer shell) so that any release from the inner tank is contained by the outer shell - 373-3.10(d)(5)(iii)(<u>a</u>);

(b) ⸻ protected, if constructed of metal, from both corrosion of the primary tank interior and the external surface of the outer shell - 373-3.10(d)(5)(iii)(<u>b</u>); and

(c) ⸻ provided with a built-in, continuous leak detection system capable of detecting a release within 24 hours or at the earliest practicable time, if the owner or operator can demonstrate to the commissioner, and the commissioner concurs, that the existing leak detection technology or site conditions will not allow detection of a release within 24 hours - 373-3.10(d)(5)(iii)(<u>c</u>).

F. Ancillary Equipment - 373-3.10(d)(6).

1. ——— Ancillary equipment must be provided with full secondary containment (e.g., trench, jacketing, double-walled piping) that meets the requirements of 373-3.10(d)(2) & (3), *unless* they are aboveground and visually inspected for leaks on a daily basis.

AIR EMISSIONS, SUBPART AA, BB, AND CC

Company Name   _____

EPA ID# No.   N _____

Region/Inspector   _____

Inspection Date   _____

Indicate:

X Violations

Indicate:

X   Satisfactory

NA   Not Applicable

Appendix X

AIR EMISSIONS-SUBPART AA, BB and CC CHECKLIST
(Only for Permitted TSDs, Interim status TSDs, and LQGs)

Subpart AA

Background: If a facility (TSD or LQG) manages hazardous wastes greater than 10 ppmw of organics in a process vent used in distillation, fractionation, solvent extraction, thin-film evaporation, air or steam stripping, Subpart AA may apply. Subpart AA would not apply in a bona fide closed loop scenario at LQGs and TSDs. To comply, the facility would need to determine if the process vent(s) releases greater than 3.0 lbs/hr or 3.1 tons/year of organic air emissions to the atmosphere. If it does not release that much then the facility is in compliance with Subpart AA. If its emissions are greater, then a control device is necessary to bring the facility into compliance. The control device may be a condenser, flare, carbon absorber, etc., that brings the equipment's emissions rate below 3.0 lbs/hr **and** 3.1 tons/year, **or** reduces the organic emissions by 95%.

Objective: The Inspector should try to determine if Subpart AA applies at a particular facility and, if applicable, evaluate the facility's efforts to achieve compliance. Has the facility calculated or measured the organic emissions from all vents and compared that with the emissions limit?

1. IDENTIFICATION OF AFFECTED PROCESS VENTS - 373-3.27(a)

   (a) Does the facility have any hazardous waste management unit using the following process? _____ Yes _____ No

   _____ Distillation

   _____ Fractionation

   _____ Thin-film evaporation

   _____ Solvent extraction

   _____ Air stripping

   _____ Steam Stripping

   (b) Are any of these units/processes exempt under the closed-loop recycle exemption? _____ Yes _____ No

   Please Explain: _____

   _____

   (c) Does the facility manage hazardous wastes greater than 10 ppmw of organics in a process vent used in above processes?

   _____ Yes. (Complete Subpart AA)

   _____ No. (Describe the information/documentation used to make the determination and collect the supporting documentation. *Proceed to the Subpart BB checklist.*)

   _____

   _____

2. STANDARDS FOR PROCESS VENTS - 373-3.27(c)

   (a) Total organic emissions from all affected process vents at the facility are below 3 lb/hr and 3.1 tons/yr. - 373-3.27(c)(1)(i):

   _____ If Yes, the calculations/analysis or performance tests are done according to 373-3.27(e). (Provide copies of the calculation and associated information - 373-3.27(c)(3).)

_____ If No, did the facility reduce the total organic emissions, by using a control device, from all affected vents at the facility by 95 weight percent: 373-3.27(c)(1)(ii). (All TSDs must have the control device in place and for LQGs by June 1999.)

3. STANDARDS FOR CLOSED-VENT SYSTEMS AND CONTROL DEVICES - 373-3.27(d)

   (a) Please explain/describe the type of control device used at the facility:

   _____

   (b) _____ The closed-vent system and control device must meet the requirements of subdivision 373-3.27(d); 373-3.27(c)(2).

   (c) _____ The owner or operator shall monitor and inspect all control devices at least each operating day to ensure proper operation - 373-3.27(d)(6).

   (d) _____ The owner or operator shall repair all detected defects as soon as practicable, but not later than 15 calendar days after the defect is detected - 373-3.27(d)(11)(iii)(a).

   (e) _____ A first attempt at repair shall be made no later than five calendar days after the defect is detected - 373-3.27(d)(11)(iii)(b).

4. RECORDKEEPING REQUIREMENTS - 373-3.27(f)

   (a) _____ Owners and operators must record the following information in the facility operating record - 373-3.27(f)(2).

   1. _____ Information and data identifying all affected process vents, annual throughput and operating hours of each affected unit, estimated emission rates for each vent and for the overall facility - 373-3.27(f)(2)(ii) ('a').

2. _____ Information and data supporting determinations of vent emissions and emission reductions achieved by control devices based on calculations or performance tests - 373-3.27(f)(2)(ii)('b').

3. _____ Design documentation and monitoring, operating and inspection information for each closed-vent system and/or control device shall be recorded and kept up to date in the facility operating record - 373-3.27(f)(3).

4. _____ Date of each control device startup and shutdown - 373-3.27(f)(3)(viii).

5. _____ The date that any leak was detected and the date of repairs - 373-3.27(f)(3)(x).

6. _____ Records of the monitoring, operating and inspection shall be maintained at least three years following the date of each occurrence, measurement, maintenance, corrective action, or record - 373-3.27(f)(4).

Subpart BB

Background: If a facility (TSD or LQG) has equipment (any valve, pump, compressor, pressure relief device, sampling connection system, flange, openended valve or line) that contacts hazardous wastes greater than 10% organics, that facility is subject to the inspection and monitoring requirements of Subpart BB. If the equipment used to transport hazardous waste with greater than 10% organics is used for less than 300 hours per year, then it is excluded from the requirements of 373-3.28(c) through 373-3.28(k) of this subpart if the equipment is identified as required in 373-3.28(0)(7)(vi).

Objective: The Inspector should determine if Subpart BB applies at a particular facility and, if applicable, evaluate the facility's leak detection and repair (LDAR) program. Does it cover all the affected equipment, what is its frequency (monthly, quarterly) and are there records of timely (<15 days) equipment repair when leaks are detected?

1. IDENTIFICATION OF AFFECTED EQUIPMENT - 373-3.28(a)

(a) Does the facility have any of the following equipment that contain or contact hazardous wastes greater than 10% organics by weights - 373-3.28(a)(2)?
_____ Yes _____ No

_____ Pumps
_____ Compressors
_____ Pressure relief devices
_____ Sampling connections
_____ Open-ended valves or lines
_____ Valves

(b) Is any of this equipment in vacuum service, which will be excluded from this requirement - 373-3.28(a)(4)?
_____ Yes _____ No

Please Explain: _____

(c) Is any of this equipment that contains or contacts hazardous waste with an organic concentration of at least 10% by weight for a period of less than 300 hours per calendar year, which will be excluded from this requirement - 373-3.28(a)(5)? _____ Yes _____ No

Please Explain: _____

(d) _____ Each piece of equipment covered under these requirements shall be marked in such a manner that it can be distinguished readily from other pieces of equipment - 373-3.28(a)(3).

(e) _____ Any equipment or device that is equipped with a closed vent system capable of capturing and transporting leakage to a control device is exempt from these requirements provided that the closed-vent systems and control devices shall comply with the provisions of subdivision 373-3.27(d); 373-3.28(k).

2. OPERATING STANDARDS:

LIGHT LIQUID SERVICE: For a hazardous waste to be in light liquid service, the vapor pressure of one or more of the organic constituents in the material must be greater than 0.3 Kilopascals at 20 degrees C and the total concentration of pure organic constituents having a vapor pressure greater than 0.3 kilopascals at 20 degrees Centigrade is equal to or greater than 20% by weight.

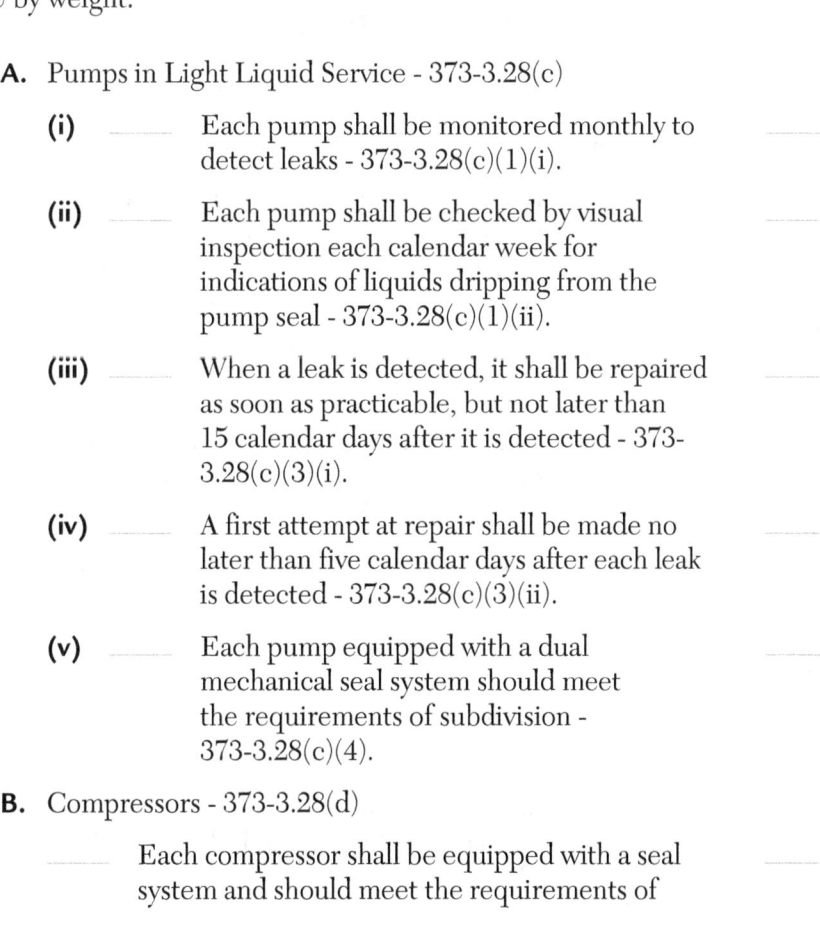

A. Pumps in Light Liquid Service - 373-3.28(c)

(i) Each pump shall be monitored monthly to detect leaks - 373-3.28(c)(1)(i).

(ii) Each pump shall be checked by visual inspection each calendar week for indications of liquids dripping from the pump seal - 373-3.28(c)(1)(ii).

(iii) When a leak is detected, it shall be repaired as soon as practicable, but not later than 15 calendar days after it is detected - 373-3.28(c)(3)(i).

(iv) A first attempt at repair shall be made no later than five calendar days after each leak is detected - 373-3.28(c)(3)(ii).

(v) Each pump equipped with a dual mechanical seal system should meet the requirements of subdivision - 373-3.28(c)(4).

B. Compressors - 373-3.28(d)

Each compressor shall be equipped with a seal system and should meet the requirements of subdivision 373-3.28(d)(1) thru (9) - {i.e. daily inspection and implementation of leak detection and repair (LDAR) program}.

C. Pressure Relief Devices in Gas/Vapor Service - 373-3.28(e)

(i)   Except during pressure releases, each pressure relief device shall be operated with no detectable emissions - 373-3.28(e)(1).

(ii)   No later than five calendar days after each pressure release, the device shall be monitored to confirm the condition of no detectable emissions, as indicated by an instrument reading of less than 500 ppm above background - 337-3.28(e)(2).

D. Sampling Connections - 373-3.28(f)

    Each sampling connection shall be equipped with a closed-purge, closed-loop, or closed-vent system and shall meet the requirements of subdivision - 373-3.28(f)(1) thru (3).

E. Open-ended Valves or Lines - 373-3.28(g)

    Each open-ended valve or line shall be equipped with a cap, blind flange, plug or a second valve and shall meet the requirements of subdivision - 373-3.28(g)(1), (2) & (3).

F. Valves in Gas/Vapor Service or in Light Liquid Service - 373- 3.28(h)

(i)   Each valve shall be monitored monthly to detect leaks by specified methods - 373-3.28(h)(1).

(ii)   When a leak is detected, it shall be repaired as soon as practicable, but no later than 15 calendar days after it is detected - 373-3.28(h)(4)(i).

(iii)   A first attempt at repair shall be made no later than five calendar days after each leak is detected - 373-3.28(h)(4)(ii).

G. Pump and Valves in Heavy Liquid Service, Pressure Relief Device in Light or Heavy Liquid Service, and Flanges and Other Connectors - 373-3.28(i)

    (i)   Pumps and valves in heavy liquid service, pressure relief devices in light or heavy liquid service, and flanges and other connectors shall be monitored within five days by specified methods, if evidence of a potential leak is found by visual, audible, olfactory, or any other detection method - 373-3.28(i)(1).

    (ii)   When a leak is detected, it shall be repaired as specified in this subdivision - 373-3.28(i)(3) (i.e., 15 days to repair and five days for first attempt).

3. RECORDKEEPING REQUIREMENTS - 373-3.28('o')

    A.   The following information must be recorded in the facility operating record - 373-3.28('o')(2).

        (i)   List all equipment to which this section applies.

        (ii)   Equipment ID number and hazardous waste management unit identification.

        (iii)   Approximate locations of units within the facility.

        (iv)   Type of equipment (e.g., plum or valve).

        (v)   Percent-by-weight total organics in the hazardous waste stream at the equipment.

        (vi)   Physical state of hazardous waste at the equipment (e.g., gas/vapor or liquid).

(vii)   Method of compliance with the standard (e.g., "monthly leak detection and repair" or "equipped with dual mechanical seals").

(viii)   The date the leak was detected and the date of repairs - 373-3.28('o')(4).

Subpart CC

Overview: The Subpart CC regulations apply to large quantity generators and treatment, storage and/disposal facilities that manage hazardous waste of volatile organic concentration of 500 ppmw or more on an average annual basis in tanks and containers.

For tank storage, there are two levels that a facility may use to manage their waste. Tank Level 1 requires a fixed roof tank which uses a maximum organic vapor pressure to comply with Subpart CC. Tank Level 2 designs can be one of five options. These are: (1) an Internal Floating Roof (2) an External Floating Roof (3) a tank with a Fixed Roof vented through a closed-vent system to a control device (4) a Pressure Tank (5) a tank located inside an enclosure that is vented through a closed-vent system to an enclosed combustion device.

Most of the facilities will comply with Tank Level 1 which is the easiest to follow. The other option that will be seen a lot would be Tank Level 2 Option 3. The other options will be limited to a small number of facilities.

For container storage, most of the facilities will store their waste in DOT approved containers. RCRA regulations already cover such storage and, as a result, most facilities will be in compliance with the container storage regulations of the Subpart CC regulations.

1. IDENTIFICATION AND APPLICABILITY:

     A. Does the facility have any of the following units that treat, store or dispose of hazardous waste with volatile organic (VO) concentrations of 500 ppmw or more on an average annual basis?
         _____ Yes _____ No